USEFUL FACTS ABOUT ABOUT ENGINEERING

JOHN STONER

LIGHT SWITCH
PRESS

Published by:

Light Switch Press

PO Box 19

Livermore, CO 80536

INTRODUCTION

I hope you enjoy this book.

I have tried to cover all aspects of engineering in an informative and easy to follow way that I hope will appeal to all of you.

Engineering is science made real and when the past comes to life. It cannot occur in isolation for engineering to be a worthwhile task, and there must be a positive economic impact to satisfy the time, effort, and funding invested.

There has never been a more exciting time to be an engineer than the present. There are now more extensive and more expensive engineering projects than ever before. We will consider all engineering disciplines during the alphabet trail from A to Z.

The author appreciates that some people will not have access to a computer, so I have tried to compensate for this fact. For the rest of the readers, the Google keyword search term is found before the brackets (the readers can find out more for themselves).

Happy reading!

CHAPTER ONE – THE LETTER A

There is no better way to start the Letter A than with aeronautics. Since childhood, I have been fascinated with flying machines, and I remember going to air shows like Greenham Common International Air Tattoo during my teenage years and thoroughly enjoying the experience. This chapter will not simply look at aircraft history but the broader engineering aspects.

The history of aircraft has been well documented over the years. Therefore I will only cover a brief history, and the reader can find out more for themselves. The history of aircraft starts with balloons! Joseph-Michel Montgolfier (1740-1810) and Jacques-Etienne Montgolfier (1745-99) filled a balloon with hot air that flew. This balloon was seen in the recent programme Around the World in Eighty Days (starring David Tennant). Gradually, other means of air flight took shape. It was a British engineer, George Cayley (1773-1857), who in 1853 built the world's first real glider. The test pilot was his terrified servant, though this remains unproven. Sir George Cayley was a pioneer of aeronautics, discovering the four forces that act on aircraft. These are weight, lift, drag, and thrust.

When we consider the concept of weight, it is the force of gravity trying to pull the aircraft downwards out of the sky. Gravity is a constant, and we will discuss constants in a later chapter. Lift happens when air travels over the wings of an aircraft. The design concerns an air-foil, a tapered end of the wing where the air travels faster at the top than the bottom. The difference in air pressure travelling over the wing causes the aircraft to ascend into the sky. This theory is called the Bernoulli Principle (the conservation of energy) after Daniel Bernoulli (1700-82). He was an inventor whose seminal work on hydro-dynamics changed the world we knew at that time. Drag is another force that affects flight. When the aircraft is in flight, the velocity or speed of the aircraft can be affected by force known as drag. The amount of drag is three times the square of the aircraft's velocity. In other words, if the aircraft is travelling at the speed of the sound, then drag is three times that number. There are many different types of drag, and the reader should find out more themselves. Thrust is the force when a plane is pulled forward by propellers or jet engines (discussed later). No structure or animal can fly without thrust (think Chicken Run, says the author, smiling).

The next significant step in the evolution of aircraft was the airship. This aircraft was made famous by Count Ferdinand von Zeppelin (1838-1917). The airship grew larger when von Zeppelin realised that a rigid outer frame would mean bigger aircraft. When the Zeppelins first came over on bombing raids during World War One, the ugly side of aeronautical engineering was recognised. That dropping bombs on targets could cause more widespread damage at a greater distance than conventional weapons like cannons. However, due to the inaccuracy of the bombing raids, these aircraft came to be known as "baby killers" due to their indiscriminate bombing of targets. In order to track these aircraft, there was an extraordinary piece of civil engineering took place. The Denge Sound Mirrors in Kent are parabolas of a concrete structure. The largest of these parabolas' "ears" was sixty-five feet high and two hundred feet long. By installing a microphone on the front of these installations, an operator could listen to the sound wave from an approaching enemy aircraft. An aircraft travelling at eighty miles per hour could be detected up to 20 miles away. The golden age of these mirrors was the First World War. These Sound Mirrors were a form of sound detection and were the brainchild of a physicist called Percy Rothwell (1895-1981) and Major William Sansome Tucker (1877-1955). These parabolic "ears" were the forerunner of radar installations. However, by the time this defence system went fully operational, technology had practically made them obsolete. Aircraft were travelling too fast to be detected in time to stop them.

In 1935, an American pilot called Wiley Post (1898-1935) made the first solo round the world flight. This flight in 1933 took eight days in total, refuelling along the way. The problem of refuelling was unsolvable until 1934, when Algene Earl Key (1905-1976) (a DSC Distinguished Service Cross recipient) and his brother Frederick Maurice Key (1909-1971), along with their mechanic A D Hunter, solved the problem. The system is called the "Probe and Drogue System", and I will let the reader find out more. The city of Meridian, Mississippi, named their new airport Key Field in their honour. Another point to note is that Wiley Post (mentioned above) was instrumental in inventing the world's first pressure suit. One of the problems faced by Wiley Post was that his cabin was not pressurised, so high altitude long-distance flight was not possible. After many attempts, Wiley Post eventually managed a flight in 1935 at 20,300 feet (6,200 metres) above Earth with a pressurised suit. At this height, a new factor was discovered, the Jet Stream. A high-velocity current of air in the atmosphere enables aircraft to travel faster than at ground level due to the difference in air temperature. The Jet stream travels at 110 miles per hour (the reader should discover more for themselves). In the 1930s, a new aircraft engine called the four-stroke piston engine was introduced. This engine was based on the principle "Suck, Squeeze, Bang, Flow" (which crudely sounds like a good Saturday night! - I am only saying what you are all thinking). An air-fuel mixture is sucked into a chamber, where a piston squeezes the fuel at the top of the chamber. This mixture ignites with a bang, and the piston is pushed back down to its starting position. The flow is the exhaust, where the spent fuel is ejected, ready to start the system again. This action feeds energy to a crankshaft that forces the machine forward (a simplistic explanation). The most famous engine is the Rolls-Royce Merlin engine, a V-12 piston aero engine installed on the Hawker Hurricane and the Supermarine Spitfire. These aircraft were both used during the Battle of Britain. Another aircraft was pivotal in the Second World War, and this was the Avro Lancaster bomber. These aircraft helped the Allies win the Second World War and were used in the famous Operation Chastise (more on later). That is a very brief history of aeroplanes.

The next part of aeronautics that should be discussed is passenger aircraft. Post World War Two, many people wanted to travel abroad, and gradually, fares became cheaper. The first plane that I wanted to discuss is the de Havilland Comet, which was the world's first commercial jet airliner. The de Havilland Aircraft Company was established in 1920 and named after Captain Sir Geoffrey de Havilland (1882-1965). However, there were problems from the start. There was metal fatigue in the airframe and stress around the square windows. Let us look at the issue of metal fatigue (in brief). In simple terms, a metal will perform adequately till a crack forms, then the crack gets more significant, and then the metal snaps. Metal fatigue happens inside the material, which can be challenging to spot. In the case of the de Havilland aircraft, this meant that the planes broke apart in mid-air. Two design problems had to be solved. The square window design was a disaster waiting to happen! Stress fractures occurred around the windows' sharp edges, caused by the difference in pressure between the outside air and the internal air of the cabin. As the plane ascends, the air inside the cabin is pressurised as the outside air pressure decreases. Over a long time, cracks will appear in square windows due to the temperature and air pressure difference. The other major problem with the de Havilland Comet was riveting, leading to metal fatigue. The reader should find out more for themselves about metal fatigue. The failure of the de Havilland Comet meant that the aircraft manufacturers such as Boeing would dominate passenger aircraft for the foreseeable future.

There have been several aircraft disasters that have led sadly to the loss of human life. However, there is one silver lining. Safety measures have been introduced following these sad events, which have changed our world for the better. Delta Air Lines Flight 191 was a regular shuttle service for passengers between Fort Lauderdale in Florida and Fort Worth in Texas. Unfortunately, this flight crashed on August 2, 1985, killing all passengers (I will let the reader find out more for themselves). The cause of the crash was wind shear from a nearby thunderstorm, which changes the velocity and direction of wind over a short distance. After this plane crash, wind shear detection systems were placed on all aircraft. Another unfortunate plane crash that altered aeronautics forever was United Airlines Flight 232. A regularly scheduled flight from Denver to Chicago. A total loss of hydraulic fluid was caused when one of the central metal components, a titanium fan, disintegrated. After this tragic event, hydraulic fuses were used on all aircraft. One part of the smoking ban

(I am now an ex-smoker) was the ban on smoking on aircraft (yes, there was a time when you could smoke on aircraft!). Air Canada Flight 797, from Dallas in Texas to Montreal in Canada, was a regular service. On June 2 1983, there was a fire in the aircraft's toilets (probably a discarded cigarette) that burnt through electrical cables at the back of the lavatory. These melted cables meant the instruments on the control panel malfunctioned. When the plane landed, the fire and smoke in the cabin were increased by introducing oxygen when the doors were opened. After this tragedy, smoke detectors were placed in lavatories, and firefighting techniques were improved. The penultimate air accident was United Airlines Flight 173, from New York to Oregon. On the approach to landing, there was perceived to be an issue with the landing gear, as the normal warning light did not go green on the control panel. The plane ran out of fuel, causing a crash while dealing with the issue. The problem with the landing gear was due to corrosion. However, there was a lack of awareness regarding the fuel situation, so a new form of training called Crew Resource Management (CRM) was implemented. The CRM meant better communication and leadership between crew members to reduce the human error factor in air accidents. The last air accident that changed passenger aircraft flying was the 1956 Grand Union mid-air collision. This accident was between a United Airlines flight and a Trans World Airlines flight when they hit each other over the Grand Canyon. At the time of the accident, there was uncontrolled airspace, which meant that there was no air traffic control system (radar) that covered the whole United States. After this tragic accident, an aircraft control system was implemented.

Lastly, there are some innovative features that I would like to discuss that are common to most aircraft. The first person that I want to discuss is Virginius E Clark (1886-1948), who invented the Clark Y airfoil, a wing design that assisted the lift-to-drag ratio described earlier. Flown by Charles Lindberg, the Spirit of St Louis used this airfoil. The other contribution to aeronautics made by Virginius Clark was to invent Duramold, a form of moulded construction of birch and plastic resin that was lightweight and durable. Duramold was the forerunner of the aluminium and glass fibre bodies that we know today. In researching this chapter, there are two people that I have found to have laid claim to the retractable landing gear. The first person was Matthew B. Sellers,

II (1869-1932), who patented the design in 1911. The other inventor was Leroy Grumman (1895-1982), who also patented the folding wing design (useful for aeroplanes on aircraft carriers). The retractable landing gear on passenger and military aircraft is such a standard feature that we forget this was not always the case. The concept of the retractable landing gear was introduced mainly to solve the issue of drag which costs money in terms of fuel efficiency. However, there was an honest debate as the mechanics associated with the retractable landing gear were expensive and heavy against the aeroplane's aerodynamic performance. The problem was solved when aircraft became faster, which led us to our next person of interest. Sir Frank Whittle (1907-1996) invented the turbojet engine. A turbo engine increases the internal combustion engine efficiency by forcing extra air into the combustion chamber that provides the power to the engine. This engine was used in the Gloster Meteor, the first British jet fighter, and was the only jet engine used during World War Two (to bring down the V1 rockets, mentioned later). The no 616 Squadron of the RAF that used Gloster Meteors had a famous flying ace, called Group Captain Sir Douglas Bader (1910-1982, the reader can find out more for themselves). The aeronautical and systems engineer Clarence Leonard "Kelly" Johnson (1910-90) was responsible for producing designs for fighter aircraft capable of breaking the sound barrier, travelling at Mach 2 (twice the speed of the sound barrier and even Mach 3 (three times the speed of the sound barrier at 2,455mph). There are over forty planes designed by "Kelly" Johnson. His invention of the afterburner blazed the trail for supersonic aircraft (the reader can find out more for themselves). The plane to break the Mach 3 barrier was the Lockheed SR-71 Blackbird, a long-range strategic reconnaissance aircraft. The combination of a high-speed aircraft and incredibly high altitudes (Mach 3 and over 85,000 feet high) made the plane practically invulnerable. It still holds the record for the world's fastest aircraft. The next engineer mentioned is Jack Northrop (1895-1981), whose contribution to aviation was the flying wing aircraft, aeroplanes designed without a tail. An example is the Northrop B-2 Spirit stealth bomber. This aeroplane was unique in the principle of de-flecting or absorbing radar signals, making the plane undetectable. The plane is still in service and can deploy conventional and thermonuclear weapons (commonly known as H-bombs or hydrogen bombs). There is one last point; in fact, it is a question – why do aircraft now fly in a V-formation? I will answer that question later (if you cannot wait, the answer is in chapter V).

I hope this chapter has proved of interest to the reader. Man has always wanted to fly, and I hope that this chapter breeds interest in any aspiring engineer.

CHAPTER TWO – THE LETTER B

The following Letter considered is letter B. The object that comes to mind immediately is bridges. A bridge is a structural marvel that can be used to overcome any natural obstruction, such as a valley or river. From the very beginning, engineers have been trying to overcome nature for the future progress of civilisation.

Firstly, there are some standard features of different types of bridges. The deck, abutment, pile, and piers all feature in bridge construction. The deck is a fundamental part of any bridge that allows vehicles, goods, and people to pass from one side to another. An abutment is a support provided at the two ends of a bridge. The pile referred to is a pile foundation, which is the foundation commonly used in bridge construction. In order to find the hard soil layer which will make the structure stable, the pile foundation is extended to a depth of many metres (the Jiaxing-Shaoxing Sea Bridge has a foundation depth of 111 metres, equivalent to 364 feet). This piling feature means we must include the term "caisson", which means a large box. A deep piled foundation must be dug when a bridge is constructed over water. A classic method was a large box called a caisson sunk into the river. This caisson is a water-sealed box, which means that when the water is pumped out, there is a dry area for the foundation work to start. When the pier on top of the foundation is completed (explained below), the caisson is removed, and the river can flow naturally. The engineer Thomas Bouch (1822-80) developed the caisson form of engineering and the railway hinge for loading trains onto ferries. This new hinge is known by the acronym RORO (roll-on, roll-off), which means that cargo can roll onto a ship rather than be craned on (LOLO = lift-on, lift-off). I will let the reader find out for themselves. It is a sad note that Thomas Bouch was the engineer held responsible for the Tay Bridge Disaster (more on later). A pier is the solid part of the bridge on top of a piled foundation that supports the bridge's deck. There are two functions of a bridge pier. Firstly, the bridge's weight (or vertical load) is transferred to the foundation. Secondly, horizontal forces act on the bridge, such as winds and river currents that make the bridge move. The span of a bridge is the distance between piers.

Some common factors affect all bridges, so let us start before considering the actual structures. One of the main factors that all bridges have to deal with is temperature. The Samuel de Champlain Bridge in Montreal, Canada is a cable-stayed bridge (please see below). The temperature range in Montreal is between 35OC and minus 30OC. All students should note that metal expands and contracts; in this instance, the metal in the bridge expands and contracts up to three metres in length! Therefore, flexible bearings are introduced to allow for the movement of the structure. The next factor that every bridge (mainly in the northern hemisphere) has to deal with is snow and ice. In order to stop the bridge from freezing, sodium chloride (salt) is used in significant quantities to keep traffic flowing. However, salt corrodes metal. There are hundreds of different methods to prevent corrosion, including painting the steel (the Firth road bridge was a prime example!). There is another factor that the engineering student should be aware of, and that is seismic activity. The South Rangiteiki Viaduct is a box girder bridge on North Island. The bridge is 315 metres long, with six spans and over seventy-eight metres high. New Zealand is prone to a large amount of seismic activity. Therefore, dampers were fitted into the structure's base so that the bridge could move but not collapse. This factor was the work of Dr Ivan Skinner (1923-2014), a leader in the field of seismic activity. Many new bridges are fitted with sensors regarding seismic activity, temperature, and many other factors.

There are, in fact, several different types of bridges. Examples include beam bridges, floating bridges, truss bridges, arch bridges, cantilever (and bascule) bridges, cable-stay bridges, and suspension bridges. We will now look at examples of these in turn. The oldest type of bridge is a beam bridge, which has long beams of wood, stone, or metal supported on either end by piers. However, there is an issue if the gap between two ends is long, as the beam bridge will only work over short distances as a single span. The Terland Klopp Bridge was built around 1800 across the river Gyaån in Norway. It is an example of stone slab bridges, known as clapper bridges, and it is considered the longest stone slab bridge in Scandinavia at 60 meters and 21 pylons. However, a beam bridge can be joined to another beam bridge, with a pier in the middle, and this will support a bridge over longer distances. This type of bridge is known as a continuous span bridge. The Feiyunjiang Bridge in China is an example of this continuous span beam bridge (I will let the reader find out more for themselves). The next type of bridge I want to discuss is the floating bridge. This bridge type is unique, as there are no foundations! These are also known as pontoon bridges, normally temporary structures (sometimes used in warfare). One of the first permanent structures was the Queen Emma Bridge in Curacao, an island in the Caribbean Sea. The example that I want to use is the Evergreen Point Floating Bridge in Seattle, Washington. The bridge is 7,710 feet (2,350 metres) long, making this bridge the longest floating bridge globally. The load is supported by floats called pontoons, which dictate the amount of traffic that can cross the bridge. There are seventy-seven pre-cast and pre-stressed concrete pontoons anchored to the bottom of the lake. The bridge's deck sits on over seven hundred concrete sections elevated over six metres above the pontoons. I will let the reader find out more. The next bridge that I want to discuss is the truss bridge. A truss bridge has straight beams that interlock in a triangular pattern, a lattice pattern of beams, and is mainly used for railway traffic. The beams act in a coordinated fashion, with tension and compression forces acting on the beams. There are no bending moments allowed in the structure. The Bollman Truss Semi-Suspension Bridge in Maryland, the United States, is a wrought iron bridge constructed in 1852. This bridge was the first all-metal railroad bridge in the world. It is now a National Historic Landmark. There are many different truss bridges, and the reader can find out more. The Kansai International Airport Access Bridge is the longest double-decked truss bridge globally. The bridge carries six lanes of automotive traffic on top and two railway lines below, over nine truss spans. The next bridge type that I want to discuss is the arch bridge. One element defines these bridges; the deck is supported by an arch below. The arch is a curved structure spanning two ends and may support a load from above. The arch works in compression, transferring the load from above into the ground below. There may be supports on either side of the arch called abutments (as they are next to or abut the arch). Viaducts and aqueducts, a type of bridge, comprise several small spans that cross a valley or river. These can also be arch in design, and the Aqueduct of Segovia is a Roman bridge in Spain. The Westminster Bridge in London, which connects Lambeth to Westminster, is an arch bridge with an interesting history. The Church of England collected taxes from the ferrymen that plied their trade across the River Thames. These taxes were big business, and the Church owned the horse ferry trade across the river. Permission to build the bridge finally received Royal Approval on May 20 1736, when George II was on the throne, and work was finally completed in 1750. The watermen received £25,000 in compensation for the loss of business, and the Archbishop of Canterbury went away with £21,025 a large sum of money. Sadly, within 100 years, the bridge was in dire straits and showing signs of subsidence. The bridge was replaced with a new bridge the one still in use today opened on May 24 1862. An example of an arch bridge that the reader might like to investigate is the Sydney Harbour Bridge in Sydney, Australia. This bridge has a fascinating history. Another arch bridge is the Garabit Viaduct in France. The designer of that bridge was Gustav Eiffel, who was also famous for the Eiffel Tower in Paris, France. The oldest arch bridge is the Arkadiko bridge in Greece opened around 1100 BC! I will let the reader find out more. The next type of bridge is called a cantilever bridge. These bridges work by being supported only on one end. The engineering principles work the same as a normal cantilever, with tension and compression working in harmony. Cantilevers are mentioned in another book, Pythagoras in the Corner. This book describes the balconies built by Francis Matcham in his theatres (the reader can find out more for themselves). The standard cantilever bridges jut out into the water and lift when river traffic needs to pass underneath. The Forth Bridge, between South Queensferry and North Queensferry in Scotland, is the most famous example. The bridge was completed in 1890 and was awarded UNESCO World Heritage Status in 2015. The bridge has been adopted as a symbol of Scotland. The bascule bridge is similar but works on a different principle. The bascule bridge is known as a lifting bridge, with a counterweight that balances the span of the bridge. Bascule means a balance scale, so we can see immediately how the concept works. The most famous example of this type of bridge is Tower Bridge in London. This bridge is a combined bascule and suspension bridge (mentioned below). The bridge was completed in 1894 and is 244 metres (over eight hundred feet) long. There are two iconic bridge towers, which have walkways between them. The bascule and lifting machinery is at the base of each tower. Each bascule weighs over a thousand tons each, and today there is an electro-hydraulic system for the lifting and lowering of the deck (the reader can find out more for themselves). The next bridge type is the cable-stay bridge. This type of bridge has cables attached directly to the pylons and secured to the roadway. The Albert Bridge connects Chelsea to Battersea on the other side of the River Thames. This bridge is the first cable-stay bridge in the world. This bridge is the only bridge that has never been replaced, apart from Tower Bridge. The bridge was opened in 1873 by designers Rowland Mason Ordish (1824-86, Royal Albert Hall, St Pancras Railway Station) and later Joseph Bazalgette (1819-91, famous for the London sewer system in the London Embankment). There were issues with the bridge as soon as it was constructed. The bridge was nicknamed "The Trembling Lady", and troops crossing the bridge had to break step (not in formation) due to the vibrations caused. The main problem with the bridge was the material used; the iron rods used in the structure suffered corrosion (which could lead to metal fatigue). The iron chains holding the deck were replaced with steel. Additional piers were added to the structure due to the advent of the motor car, which the bridge could not cope with the extra weight. There are still traffic restrictions to this day. A cable-stayed bridge has one or two towers (called pylons). These towers have cables extending from them that support the bridge's deck (which supports the traffic going over the bridge). This bridge is different from a suspension bridge (mentioned later) due to the cantilever design of the bridge itself (cantilevers mentioned earlier). The longest cable-stayed bridge globally is the Jiaxing-Shaoxing Sea Bridge, which opened in 2013 in China. The most famous example is the Millau Viaduct Bridge, which opened in 2004 in France. However, the bridge that I want to mention now is the Rio-Antirrio Bridge (mentioned in the programme "Impossible

Engineering" on the Yesterday channel.) This bridge is located in the Gulf of
Corinth in Greece and opened on August 12 2004. Several special considerations were taken into account during the design of the bridge. The foundations
were over six and a half metres deep, as the seabed was all loose sediment.
There were iron rods piled into the seafloor, and the foundations laid on gravel.
This action meant that the foundations could move in the event of a seismic
shift, which the Gulf of Corinth is prone to regularly having. A similar arrange-
ment can be seen in the Guy Maunsell (1884-1961) naval forts that are still
present in the Thames estuary. Another special consideration was the use of
dampers; these are connected to the bridge to stop any unwanted movement on
the bridge. This action happened to the London Millennium Bridge (opened in
June 2000), where dampers were installed to stop the bridge's sway. Other
considerations have been covered earlier in the chapter. A bridge that is worthy
of mention is the Oresund Bridge, which connects Copenhagen in Denmark to
Malmo in Sweden. The bridge is eight kilometres (five miles) long from Sweden to an artificial island called Peberholm in the Oresund Strait. The remainder
of the journey is underground, by the Brogden Tunnel from Peberholm to the
Danish island of Amager. This bridge has been included as it forms the backdrop to the crime series The Bridge, which started in the United Kingdom in
2012. Of course, I would recommend the reader enjoy this amazing series and
the bridge itself. The last type of bridge is the suspension bridge. These bridges have a deck suspended in the air on cables that are on top of pylons (towers).
There are anchorages on both sides of the bridge. The most famous suspension
bridge, according to the author, has to be the Clifton Suspension Bridge. This
bridge spans the River Avon in Bristol, and the designer responsible was Isambard Kingdom Brunel (1806-59), one of England's greatest engineers! This
bridge has a fascinating history, beset by financial problems that ultimately
affected the end design. It is a sad footnote that Isambard Brunel never lived to
see the bridge completed, having died five years before the work was finally
completed (I will let the reader find out more for themselves).

The last part of this chapter concerns abandoned bridges. There is a
fascinating history of abandoned bridges, and I will mention some of these
now. These bridges have been featured in various television programmes, and
I will only give a brief history here. The first abandoned bridge is nicknamed

"Hitlers Bridge", also known as Borovsko Bridge in the Czech Republic. Construction of this bridge started in 1939, at the beginning of the Second World War. However, with the assassination of Reinhard Heydrich (Operation Anthropoid, the reader can find out more for themselves), and the invasion of Soviet Russia by Germany in 1941, work was halted. The work on the bridge started again in the 1950s; however, work stopped again. The whole area around the bridge is now shut to the public. I have included this bridge as it is a testament to the failure of Nazi Germany to change the world. Another bridge stands as a testament to the failure of Nazi Germany, and that is the "Bridge to Nowhere" in Latvia (the reader can find out more for themselves). Another bridge is the Goat Canyon Trestle Bridge in California, United States. This bridge is the largest wooden trestle bridge in the United States. It was built of wood in the 1930s due to the high-temperature fluctuations in the desert, rather than steel which would deteriorate quicker. However, the railroad company never turned an adequate profit, and the railway line was eventually closed. The bridge is still maintained; however, the author's opinion is that the bridge may be used again. However, it costs a fortune in maintenance, so the future remains uncertain. The next bridge I want to mention is the Kinzua Viaduct Bridge, built-in 1882, located in Pennsylvania, United States. This bridge was the fourth tallest bridge in the United States. The tornado that finished the bridge arrived in July 2003, with ninety-four miles per hour winds. This wind tore the bridge to shreds as there were corroded wrought iron foundation bolts. Repair work was due to start four days after the tornado! One last bridge has to be mentioned, the remnants of which can be seen adjacent to the Firth of Forth Bridge mentioned earlier. The story starts with a disaster that spelt the end of a career, some say. The Tay Bridge disaster in 1879 was one of the worst rail disasters in United Kingdom history. The engineer Thomas Bouch (1822-80) was an outstanding engineer whose achievements we have discussed earlier. With the subsequent loss of life, the Tay Bridge collapse spelt the end of his career. On a personal level, Thomas Bouch never fully recovered from this disaster.

Dear reader, I have tried to cover as many bases as possible. There will always be better examples of the bridge types that I have chosen in the future.

I wanted to help the reader, and I hope I have succeeded. Thank you for your time and effort.

CHAPTER THREE – THE LETTER C

Several options were chosen for the letter C. There are always constants used in engineering equations. There is another option that I have chosen for the letter C, and that is a cantilever. These are present in many structures around the world. The other option for C has to be concrete, a material that is one of the most widely used worldwide.

The first option that I have chosen is constants. A constant in any equation is always a known factor. Therefore, I will start at the beginning and end ad infinitum (my apologies, I could not resist!). I will not discuss all of these in great detail but will try and give practical examples to show why these constants are essential. One of the first constants that I ever learned in school was Pi, which was used to discover the ratio of a circle's circumference to its diameter. Although not strictly accurate, Pi is a constant number, a fraction that we know as twenty-two divided by seven, i.e. 3.14159. A Chinese astronomer named Zu Chongzi (born 429 bc and died 500 bc) gave several close approximations of the figure of Pi. Therefore the area of a circle is Pi x radius squared. The letter Pi is used to estimate the diameter of a circle. Many constants work on the principle of proportionality; an example is that the circumference of a circle is proportional to the circle's diameter, represented by the symbol pi, 3.142 equivalent. In the Introduction to Pythagoras in the Corner, I discussed that Joseph Bazalgette (1819–91) engineered the first sewerage system in London. In order to determine the diameter of the pipe required, several calculations had to take place. There was a statistical calculation looking at the population growth expected in the city and a calculation of the effluent to be produced. Once a route had been established, Joseph Bazalgette had the brilliant idea of doubling the diameter to account for population growth. This action meant a far larger tunnel; however, this pipe proved helpful for the next hundred years plus. However, there is now double the number of population living in London that was ever envisaged when the tunnel was first constructed. Therefore a new system called the Thames Tideway Tunnel, and the Lee Tunnel have to be constructed. There is another example. I once travelled to the island of Crete, near mainland Greece. There was a civilisation called the Minoans, whose capital was called Knossos. It was the first civilisation to use underground clay pipes to carry water to their homes and the disposal of sewage. They were the first people who understood the importance of the correct diameter for their pipework! Let us consider another constant. Acceleration due to gravity, or gravity for short, is a constant. This constant is known as the letter g, and the rate of gravity is 9.81 metres per second. The gravity of the Moon affects the tides of the sea and gives weight to living objects. Let us consider this in detail by looking back into history to understand gravity fully. There were several false starts, but Galileo Galilei (1564-1642, of telescope fame), or simply Galileo, had the correct hypothesis. By dropping steel balls from the Leaning Tower of Pisa and rolling them down various hills, he realised that all objects, regardless of their size, fall at the same speed. The resistance of the air or air movement changed the figures. This theory was the starting point for the understanding of gravity. Sir Isaac Newton (1642-1726) carried this theory further and looked to the stars. Sir Isaac Newton discovered gravity by watching an apple fall from a tree (fact, not falling on his head). Sir Isaac Newton wanted to understand how the universe worked and, to put it into simple terms, how the planets stayed in their orbits around the Sun (called the inverse square law of universal gravitation). There are three laws of Newton's motion that the reader can learn more for themselves. However, there are flaws in his theories, but his contribution should not be underestimated. If you are interested, the flaw is the "wobble", for want of a better word, of the orbit of Mercury around the Sun. The orbit is not even. Newton's theory led to the discovery of Neptune as a planet. Albert Einstein (1879-1955) explained the flaw in Newton's theory with his general theory of relativity. If the reader has a computer, and I appreciate that not everyone does, this explanation can be found on Google. Other physical constants should be considered. The Kelvin scale was discovered by William Thomson, 1st Baron of Kelvin (1824-1907), whose work in physics and thermodynamics significantly accelerated humanity's knowledge of how temperature works. There are no minus temperature scales, so absolute zero in temperature is -2730C, equivalent to 00K. When discussing the insulation and thermal efficiency of buildings, all temperatures are measured on the Kelvin scale of temperature (which will be examined later in the book). On 19 July 2019, the Kelvin scale was redefined, following the Boltzmann constant (regarding the number of molecules in a gas). I will come back to this constant later in the book. Many other constants can be discovered. However, the engineering student would be well advised to discover the broader world of mathematics and understand how a constant works.

The next option I have considered for the letter C is the cantilever function. The cantilevered beam is a beam attached to a wall and supported on only one end. We had mentioned this function in the last chapter when we discussed the Firth of Forth Bridge in Scotland. The cantilevered balcony was the architect Frank Matcham's (1854-1920) brainchild, who mainly worked in theatres throughout his professional career. The galleries in Francis Matcham's theatres allowed extra seating space without additional columns for an unobstructed view of the stage. Another example is the South Bank Tower, located between Blackfriars Bridge and Waterloo Bridge in London. The building was due for demolition; however, the alternative was to make it more extensive, which made it fit for its purpose as an office building. Eleven new storeys were added to the structure, incorporating residential apartments. These new storeys were cantilevered beams from the main structure of the building. A famous house called Fallingwater by architect Frank Lloyd Wright (1867-1959). This building is one of the most iconic symbols of organic architecture and is located in Pennsylvania, United States of America. Concrete beams support the floor of the house, which is built on top of a waterfall. There are cantilevered balconies that extend from the main structure of the house. The house was designed to live in harmony with the nature surrounding the building. Some examples can be found all over the internet. These include the Beetham Tower in Manchester, England, and the Zaha Hadid Architect MAXXI museum in Rome, Italy.

The last part of this chapter concentrates on another C, called concrete, a mixture of ballast, cement, and water. Concrete is the most widely used artificial material on the planet. It is second only to water as the most widely consumed good on the planet. Research on concrete and cement (mentioned at the end of the chapter) is useful to understand. When I was a student, various tests had to be carried out on the concrete. These practice tests have their limitations; however, they may be useful for any student to understand the theory behind the test. The first test I want to discuss is the slump test, which involves filling a conical cylinder with concrete. The concrete is then tampered down with a rod till the cylinder is full. The cylinder is then turned upside down and removed. The concrete should have a slight tendency to fall over. If the concrete falls over too easy, then the concrete is too runny and will be unsuitable for bearing any loading. If the mixture is too stiff, this can also cause problems and be too brittle. The method of working out if the mixture is correct is to measure the height of the concrete. This test is based on the theory of plasticity of materials, which the reader can find out about themselves. The second practical test of concrete is the cube test, which involved taking random samples of concrete delivered to a construction site. These concrete samples were put into steel cubes, tamped down, and left in a water tank. After some time, usually seven days, the samples were delivered to a specialist laboratory. The concrete cubes are subjected to load pressures, which measure the compressive strength of the concrete material. The Law of physics that this function relates to is known as Hooke's Law, named after Sir Robert Hooke (1635-1703). Hooke's Law states, "stress is directly proportional to strain until the proportional limit is reached, at which point the object tested will begin to show signs of stress". Concrete is one of the world's most used materials. Therefore, any concrete structure with reinforcement rods shows that it can carry the intended load, such as a skyscraper. Other factors affect the performance of concrete. The porosity of concrete measures the voids created by air bubbles in the concrete to the ratio of the concrete structure itself. However, porosity in steel is usually caused at the welding stage, and the reader should investigate this matter for themselves. There is another factor that should be considered, and that is the permeability of concrete. This action measures how water flows through the voids in the concrete structure. In very high-quality concrete, the level of permeability is slow, such that it would take over four thousand years to permeate a six-inch wall of high-quality concrete. However, the engineering student should know that it is impossible to create a totally watertight concrete structure. There is another factor regarding concrete called durability that the reader should know. The durability of concrete is simply the ability to last for a long period without deterioration. The factors that affect concrete are temperature, humidity, and chemical resistance. The location of the concrete structure will define the type of concrete used. I have not gone into any detail; however, these are for your attention. Another factor that affects concrete includes fire resistance. Concrete as a material cannot be set on fire; it does not burn or give any toxic fumes due to a fire. Therefore, concrete structures are ideal when considering the fire resistance required of any building. There is one last bit of history that I would like to add to this chapter, featured on the programme "Abandoned Engineering". On the outskirts of the city of Berkeley in Gloucestershire lies the village of Purton. There is a canal called the Gloucester and Sharpness Canal, which serves the town. The River Severn is adjacent to the Canal. The River Severn has one of the highest tidal ranges in the world (the difference between low and high tide), up to fifteen metres (forty-nine feet). A bank separates the Sharpness Canal and the River Severn; however, due to the tidal range of the River Severn, the bank was gradually being eroded. On 3 December 1909, a severe storm across the southern part of England threatened to destroy the bank and lose the Canal forever. In response to this, several wooden barges were run aground to try and form a barrier to the erosion of the bank. However, this was only a temporary measure. During World War Two, there was a shortage of materials for making boats, such as timber (more on later). However, concrete barges were a solution, as these were made from both concrete and reinforced with iron. These were used during Operation Overlord, the invasion of France on 6 June 1944 (D-Day), that started the end of the Second World War. These barges were run aground at Purton to reinforce the sandbank and preserve the Gloucester and Sharpness Canal to this day (there are further examples at Rainham Marsh). I have not included all information about concrete, as that would take far too long and would only be of interest to a few individuals. However, I hope that this small introduction to concrete helps the reader to understand the significance of this commonly used building material. Concrete has one major disadvantage, not as an engineering material but in producing this material. Cement, a key ingredient of concrete, is responsible for around eight per cent of carbon dioxide emissions (greenhouse gases). Over a series of different books, I will be considering the previous statement in greater detail. However, in the meantime, there are many academic papers and articles dealing with concrete greenhouse gas emissions (the reader can find out more for themselves).

I hope the reader has found this chapter both informative and entertaining.

CHAPTER FOUR – THE LETTER D

For this chapter, dams are chosen to represent the letter D. There are so many different varieties and uses for dams that I felt they deserved a whole chapter. I would like to firstly cover the functions of dams before discussing the types of dams in existence. The chapter will conclude with dam failures, their reasons, and what lessons can be learned.

There are six functions of dams that I would like to discuss first. One of the functions of a dam is water storage and water supply. The first point to note is that rivers discharge water into the sea, and by blocking this flow, we create water reservoirs. These water reservoirs can also be created by blocking tributary rivers that feed the main river. By storing the water, fresh drinking water is supplied to the local towns. In Wales, the Claerwan reservoir and dam are used to supply fresh drinking water to the city of Birmingham and the West Midlands area. One of the most famous dams is the Hoover Dam in America; at 726 feet high and 1,244 feet long, it was one of the most significant manufactured structures in the world at its construction. The dam is used to store water, and the reservoir is called Lake Mead, which is one of the largest reservoirs in the United States. This Lake is used to supply fresh water to the city of Las Vegas (we will discuss the issues regarding water efficiency at a later point in the book). Another function of a dam is the irrigation of crops for agricultural use. The water in the dam is channelled outwards so that crops located at a long distance from the reservoir have a clean supply of water. One of the most famous examples is the Aswan Dam in Egypt, built-in

1970 to dam the River Nile. The water reservoir created is called Lake Nass-er, and now thousands of acres of Egyptian barren, arid soil can be used for agriculture. One of the most visual reasons for creating a dam is the need for hydroelectric power. A hydroelectric power plant has three parts: a reservoir where the water is stored, a dam that holds the water inside the reservoir, and an electric plant that produces electricity. Gauges on the dam keep open, and water flows through, powering the turbines that generate electricity. The power created is in Alternate Current (AC); therefore, a transformer is required to transform the electricity into a higher voltage current that can be distributed along with the power network. The largest producer of hydroelectric power is the Three Gorges Dam in China. The dam is located in Hubei province in China, and in 1994 construction started on the dam, which opened in 2011. Over twenty-eight million tons of concrete were poured over 460,000 metric tons of steel to create a dam that stands over 607 ft high and 7,575 ft long. The Three Gorges Dam has the potential to create over 22,500 megawatts of hydropower electricity. In the United Kingdom, the first central pumped-stor-age hydroelectric power station was the Ffestiniog Power Station in Wales. It was completed in 1963 and provided 360 MW of electricity to the National Grid. When power is required, the reservoir Llyn Stwlan starts to empty. The water is forced down two 200 metre shafts that deal with a pressure of 27 cubic metres of water a second. The water is forced into four concrete tunnels that feed the water through the turbines' valves. The water then proceeds onto another reservoir called Tan-y-Grisiau. The water is then pumped back up to the top reservoir, and the cycle starts again. The Dinorwig Power Station is located in Snowdonia National Park in North Wales and is a pumped-storage hydroelectric power station. The construction of the power station started in 1974 and was completed ten years later. The project's cost was £425 million and delivers 1,650 MW of power via six turbines. The power station is located adjacent to an exhausted Dinorwic slate quarry, now home to the National Slate Museum. The power station is located deep inside the mountain Elidir Fawr, with tunnels and caverns leading to the power station's heart.

Another function of a dam should be considered, and this is flood control. These dams are called flow-through dams or perforation dams. They are con-structed solely for flood control. There is a spillway built at the same height as the river bed, allowing for the natural flow of the river. When water levels are high, the dam effectively controls the amount of water flowing into the river. It should be mentioned that there is, under normal conditions, a negative impact on the environment. There is no water stored at a flow-through dam, which is not its primary purpose. There are examples of flow-through dams all around the world. In the United Kingdom, the Llyn Brianne dam regulates the water flow of the River Tywi in Wales. In the United States, there are dams in the Tennessee Valley that regulate the flow of the Mississippi River, amongst others. We have mentioned earlier the Aswan dam, which was partly built due to the annual flooding of the River Nile in Egypt. A point of interest to the reader is the twin temples of Abu Simbel, relocated in their entirety to avoid being submerged during the creation of Lake Nasser. The cost of the project was around forty million US dollars. The Three Gorges Dam in China is used to alter the flooding of the Yangtze River, though there are doubts about how effective this has been. I hope this section helps the reader understand the functions of dams; we will now consider the different types of dams there are in existence.

The first type of dam is the Gravity dam. These are dams that resist the water pressure of the reservoir behind the dam using their weight. They are generally triangular and thicker at the bottom than at the top of the structure. The Grand Coulée Dam in Washington, USA, is a gravity dam completed in 1942. There is a water reservoir behind the dam called the Franklin Delano Roosevelt Lake. The dam has been used to irrigate Banks Lake and the surrounding area. The power generated by the dam is used for industrial purposes. The second type of dam is the Arch dam, a curved dam with the convex side of the arch facing upstream. The arch dam is suitable when there are narrow canyons at your location. In Europe, the Lac de Montsalvens is a reservoir created by an arch dam in Fribourg, Switzerland. Some dams combine these two types, called a gravity-arch dam, such as the Hoover Dam, mentioned earlier. The Hoover Dam houses Lake Mead, supplying water to the city of Las Vegas. Without the Lake, Las Vegas would not survive as a city built in the Nevada desert. A buttress dam consists of a face slab, buttresses, and a base slab. These dams are used where the soil is weak or in a broad valley, and extra support is required. The student should consider the Norwegian civil engineer Nils

F. Ambursen (1876-1953), who invented the buttress dam. There are many examples of buttress dams worldwide, especially in France and the United States. An embankment dam is one of the earliest and most common dam types. Earthen embankments or rock-filled embankment dams have material piled on top of each other to create a reservoir of water. In the dam's centre, a layer is made up of impermeable material such as clay type soil or even asphalt that tops water passing through the dam. In rock-filled embankment dams, the face of the dam that faces the reservoir (the upstream face) is generally covered in concrete or asphalt. There are hundreds of examples throughout the world, and I would like to discuss a few examples now. Fisher Pond Dam in West Tisbury, Hampshire, is a type of earthen embankment dam. The dam was constructed in the early nineteenth century by Dr Fisher, who owned the land, and was for recreational purposes. According to a Report in 2009 by a civil engineer called Mr K.A. Healy, the dam is still in good condition to this day! The reader may also wish to learn about the fish ponds near Frensham Ponds; earthen embankment dams used to provide fish for the medieval diet. A neo-Gothic solid masonry dam was constructed at Derwent Valley in Derbyshire, England. This water reservoir was for the clean supply of water to the North East of England, including Nottingham and Leicester. The dam became famous as a practice ground for the Dambuster's raid (mentioned later).

The Tarbela Dam in the Haripur district in Pakistan is the largest earth-filled dam in the world. The dam was completed in 1976 and had the dual function of irrigation and flood control of the Indus River (the reader can find out more for themselves).

The next section on dams considers the failure of dams and what lessons can be learned from these mistakes. The first failure of a dam that I want to discuss is an act of deliberate sabotage. During the Second World War, the factories of the Ruhr Valley in Germany were of paramount importance to the German war effort. This area was heavily industrialised, producing steel for munitions, such as tanks and guns. The main source of power for these factories was hydroelectric power from the three dams located in the area. It was decided, by the Allies, to use the codename Operation Chastise to bomb the three dams. This operation became famous as the "Dambusters Raid", an operation to destroy three dams situated in the Ruhr Valley: the Mohne Dam, the Edersee Dam, and the Sorpe Dam. Each of these dams had walls that were over forty metres across. The solution was presented by an engineer working at Vickers, called Sir Barnes Wallis (1887–1979). His invention came from historical records, stating that Admiral Nelson used a novel method to hit enemy vessels. The cannonballs fired from his ship skipped along the water's surface and hit the target. Two forces work on an object, like a stone, skipping along the surface: gravity, which is pulling the stone downwards and lifting, the reactive force of the water that keeps the stone buoyant when it hits the water. The main key to the problem was speed. The stone must be travelling at a certain rate per hour, with enough revolutions per second to keep moving forward at a certain distance before sinking below the surface. This action is quite complicated physics, but it makes for an interesting physics lesson! Barnes Wallis realised two functions were required. An aspherical shaped bomb would be required. Backspin was needed for the bomb to sink at the end of its run, exactly where it should. By continuous testing, it was found that a nine thousand pound bomb had to be dropped from a height of thirty metres (100 feet) above the water, at speed not less than 230 miles per hour with a backspin of 500 rpm (revolutions per minute). The backspin was achieved inside the aircraft by a conveyor belt. However, there was a further problem. The planes could not have lights, as they were travelling in a straight line and were easy targets for the Germans. The problem was solved by placing lights underneath the aircraft, so when the lights met in a figure of eight patterns, the correct height was reached, and the bomb could be dropped. There were practice runs made by 617 Squadron, led by Guy Gibson, over the Derwent Reservoir in the Upper Derwent Valley in Derbyshire, and these were success-ful. On the night of 16 May 1943, the bombing raid started. It was successful, and Mohne Dam and the Edersee Dam were breached, which caused flooding in the Ruhr Valley. Two hydroelectric power stations were destroyed, and other factories and mines in the Ruhr Valley. However, there was a high price to pay. Eight aircraft were shot down, 53 airmen were killed, and three airmen were captured and taken prisoner. However, Guy Gibson, for his heroism, was awarded the Victoria Cross for bravery, and several other Air Force members were decorated as well, with Distinguished Service Orders, Distinguished Fly-ing Medals, and Conspicuous Gallantry Medals. The second dam that I would like to mention is the bursting of the Bilberry Reservoir in 1852. The Bilberry Reservoir was built at Holmfirth in Yorkshire over a stream that had not been properly diverted. This action had the effect of weakening the dam walls that held the reservoir. When the reservoir broke on 5 February 1852, over 86 million gallons of water swept through the Holme Valley and into Holmfirth. The effect was catastrophic. Buildings, cottages, animals, and people were all swept away. The flood caused the loss of over eighty-one lives, four mills, and seven bridges. It was decided that defective construction had caused the damage to the reservoir dam wall. The weather also played an important part. With the water rising in the reservoir, the water started to come over the top of the dam and started washing the walls away. Eventually, the walls cracked, and the reservoir emptied downstream. The last dam that I want to mention is the Dale Dyke Dam, located near Sheffield in the north of England. The Bradfield Reservoir was built between 1859 and 1864 to provide fresh drinking water to the increasing population of Sheffield and the surrounding area. The reservoir was over 27m deep at its lowest point and contained 712 million gallons (114 million cubic feet) of water. The reservoir was held back by the 500ft wide and 30m high Dale Dyke Dam, an embankment dam made of 400,000 cubic yards of rock, shale, and stone. The Sheffield Waterworks Company built this dam in 1853 under an Act of Parliament called the Water Company's Act of 1853. The accident happened when the dam walls burst in 1864, sending millions of gallons of water through the towns and villages, including Loxley and Hillsborough and devastated parts of Sheffield. This disaster was called the Great Sheffield Flood and was one of the worst Victorian disasters. The flood destroyed over 800 houses and eleven churches. Fifteen bridges were destroyed, and five bridges were partially destroyed. It was estimated that over 200 people died that night. The experts disagreed on how the dam collapsed, with some experts blaming slippage of the ground underneath, whilst others blamed the thickness of the dam's walls. It should be noted that on the night of the dam's collapse, a gale was blowing that would have caused problems with the dam and may have been a deciding factor in the eventual collapse. It should be noted that apart from these disasters, there are thousands of dams that have been built correctly and still function to the present time.

There are advantages to dams that have been discussed earlier. However, there are disadvantages as well. When the Three Gorges Dam was constructed, over a million people were forced to relocate elsewhere. The environmental impact can be devastating for the local wildlife and aquatic life. Fish, such as salmon that require travel upstream to spawn, will find their path blocked. However, many dams now have "fish ladders" so that the fish can swim upstream. Another problem that has been encountered is drought. The southwest part of China, where the Three Gorges Dam is based, is currently experiencing the worst drought in over 100 years due to the rise and fall of the water in the area. The most severe problem is the cost of constructing a dam. The Hoover Dam cost $49 million to build, which is about $833 million in the present day with inflation. The Three Gorges Dam cost $26 billion to build. Therefore, it can be said that although there are significant advantages to building dams, they are not a cheap solution and will impact the surrounding area.

I hope this chapter has provided an insight into a small part of the world of civil engineering. Dam construction is fascinating to the civil engineer!

CHAPTER FIVE – THE LETTER E

There was only one option for the letter E, and that was engines. These are the workhouses that power vehicles. I would like to discuss the three different categories of engines. We will discover together these engines and the historical and economic impact they have had on history. Engines would not have become so popular unless there was an increase in trade.

The first category of the engine is the thermal engine. It is called the thermal engine as heat is produced due to the mechanical actions of the engine itself. This engine type is further subdivided into internal thermal, external thermal, and reaction engines. The internal thermal engine is the most common engine, better known as the internal combustion engine. Should any aspiring engineer wish to investigate, there is a comprehensive history online. For those students without access to a computer, here is a very brief outline (I apologise if there are gaps in the knowledge). There is a concentration on petroleum engines rather than diesel engines. Rudolf Diesel (1858-1913) is the namesake and inventor of the diesel engine, for which he created the prototype and gained the first patent in 1893. He is also well known for the mysterious circumstances of his disappearance aboard a ship in the English Channel. In the next chapter, I will discuss the different types of fuel used so I will leave the diesel engine to a later chapter. In the author's opinion, the history of the internal combustion engine starts properly in the eighteenth century. In 1791, an inventor called John Barber (1734-1993) received British Patent 1833 (Obtaining and Applying Motive Power &c. A Method of Rising Inflammable Air for the Purposes of Procuring Motion and Facilitating Metallurgical Operations), the first-ever gas turbine. The method of igniting compressed gas into the operation of a paddlewheel was the first attempt to create forward motion. However, although the technology was sound, there was insufficient power to compress the gas and produce motion. A French inventor called Nicephore Niepce (1765-1833) invented the first proper internal combustion engine (and who is also credited with the invention of photography). This engine, called Pyreolophore, powered a boat in 1807 up the River Soane in France using this engine. There were several more examples of internal combustion engines, some of which worked though these were of small capacity. In 1872, an American inventor and engineer called George Bailey Brayton (1830-92) introduced the constant pressure engine. This engine worked on the principle of a liquid-filled fuel injection system. The air and fuel system ignited and passed into a power cylinder that made the forward motion. The engine was used to power the world's first submarine, the Fenian Ram (designed by John Philip Holland, 1841-1914, and was launched in 1881). Due to the constant pilot flame, there was no need for constant ignition. However, the results were limited, and a better engine designed by Nicolaus Otto (1832-91) became more commercially successful. The Otto engine paved the way for the modern internal four-stroke combustion engine, as there was a timed spark for the ignition of a fuel and air mixture. Many inventors started the modern era of engines that we know today in the years after. In the first chapter, we have mentioned the actions of the four-stroke combustion engine. The Avro Lancaster Bomber, the Spitfire, and the Hurricane all played their part in helping the Allies win the Second World War. I feel another interesting section should be added to this section on the internal combustion engine. Man has always wanted to go faster and faster, and the goal of piloting the fastest vehicle on Earth has driven technology to great heights. While I could comment on many engines, I will try and outline a brief history. The land speed record is held by a vehicle on land that achieves the highest speed averaged over two runs over a fixed distance course. (In this section, I will only deal with internal combustion engines, rockets will be mentioned in a later chapter). For those students with a computer, the information is readily available. However, I will try and assist those students without access to a computer. The first person that I want to discuss is

Henry Ford (1863-1947). There were a few drawbacks and false starts for Henry Ford, as he had left the Detroit Automobile Company in 1902 to start his passion for motor racing. Henry Ford, in 1904 won the world land speed record with a speed clocked over ninety miles per hour with a car called the Ford 999. The car had a massive four-cylinder engine with no camshaft, no suspension and a 230-pound flywheel. This win put his new company, the Ford Motor Company, in the spotlight, and the rest is history. A 2.75 mile (4.43 km) race track in Weybridge, Surrey, called Brooklands. It was opened in 1907 and had impressive thirty-degree banks on the circuit. Between 1909 and 1922, three land speed records would be broken at this location. Between 1924 and 1935, there were mainly two British drivers that held the land speed record between them. Major Sir Malcolm Campbell MBE (1885-1948) and Sir Henry Segrave (1896-1930, the first man to reach 200mph) held land and water speed records. The Segrave Trophy is named after Sir Henry Segrave, who was the first person to hold both land and water speed records at the same time. The engines used for the land speed records were all V12 engines. In the 1930s, Germany was in the grips of a depression brought on by its actions during World War One. The Nazi Party, led by Adolf Hitler, wanted to establish German superiority in automobile racing. There were early successes at various Grand Prix; however, the land speed record eluded them. The Mercedes T80 was the car set to destroy the land speed record. This astonishing piece of design was by Dr Ferdinand Porsche (1875-1951), which is now known as a high specification car company. The Mercedes T80 ran on pure alcohol (ethanol mixture) and was a six-wheeled, 44.5 litre, three thousand horsepower aero engined (Messerschmitt Me309 and Me410 aeroplanes) car, designed to break the land speed record (though this was never achieved, as the outbreak of World War Two delayed the project indefinitely). The car had a fatal flaw; at speed, the car would have lifted at the front, and the downforce at the rear of the car would have killed any racing driver in a horrific accident. The next pinnacle achieved was 400 mph in 1947 in a British racing car known as the Railton Special, driven by John Cobb (1899-1952). Donald Campbell held the last internal combustion engine land speed record in 1964 when a speed of 403 miles per hour was reached at Lake Eyre, in Australia. From this point onwards, it would be the jet engines used for the record. There is even more to discover, and I will let the reader find out more for themselves. The external thermal engine is more commonly known as the steam engine. This was a period when machines took over many of the hand-crafted jobs that shaped the economy. The steam engine would not have been so successful had it not increased production and, therefore, profit for the owners of the machines. The first invention was the Newcomen Steam Engine, invented in 1712 by Thomas Newcomen (1664-1729). This invention was used primarily to pump water out of tin mines, which was a major problem for the mining industry. It worked on a beam engine design, where water was sucked out using a vacuum. Due to the uncomplicated nature of the design, thousands of these engines were built around the United Kingdom to power factories and other major works. The major disadvantage was the inefficiency of the engine, as large amounts of coal had to be used to power the engine. A Scottish engineer superseded the Newcomen design, mechanical engineer, and chemist James Watt (1736-1819), who truly invented the steam engine. This engine was a far more efficient design, as high-pressure steam worked the main engine. This invention powered machines in factories and was also used to power the first locomotives. The principle of the steam engine improved by using high-pressure steam that powered a locomotive at Penydarren ironworks at Merthyr Tydfil in July 1812. This steam engine had, in fact, an impact not only during the Industrial Revolution but on the whole world. The engineer, Isambard Kingdom Brunel, has been mentioned previously in this book. However, the reader should be aware that, apart from the Clifton Suspension Bridge in Bristol, Isambard Brunel was responsible for many bridges, tunnels and railway lines, including the Great Western Railway. Isambard Kingdom Brunel built and designed three ships; these were the SS Great Western, the SS Great Eastern, and the SS Great Britain (I will be discussing maritime engineering later). In the foreword to this book, I have mentioned that engineering does not happen in isolation; there has to be a positive economic impact. Steam engines were used in different ways. The production of goods such as cotton, silk, and other linen, would have remained a small-scale industry if steam engines had not been invented. By using economies of scale, the production increased, allowing trade to develop worldwide (this is true of other products as well). The movement of goods and raw materials was increased by introducing steam locomotive engines. The first public steam railway in the world was the Stockton and Darlington Railway, which opened in 1825. Through the advent of steam locomotives, continents such as America could finally become unified. Many steam engines have become famous over the years, so I will only mention a couple now. I have mentioned the land speed record already in this chapter. The most famous steam train globally has to be the LNER (London and North East Railway) Class A3 A472, the Flying Scotsman. This train was the first steam train to travel over 100mph, which now resides at the National Railway Museum in York. However, the LNER Class A4 A468 locomotive known as the Mallard holds the record for the fastest steam locomotive. The record of 126 mph reached on 3 July 1938 has never been beaten. The last thermal engine I will mention is the reaction engine, more commonly known as the jet engine. The engine produces thrust (going forwards) by expelling reaction mass (such as fuel) from the exhaust. This process follows Isaac Newton's (mentioned earlier) third law of motion "for every action, there is an equal but opposite reaction force". However, engineering students should know that not all reaction engines are 100% efficient. The problem is known as propulsion efficiency, and the reader can find out more for themselves. Some of the rocket engines featured are only 60-70% efficient, as some of the energy is lost as heat out of the exhaust. There is another effect called the Oberth Effect that affects the engine's efficiency. Rockets work best when they are moving, and there is a simple equation that force multiplied by distance is the definition of mechanical energy. Therefore, the faster the rocket's speed, the greater the kinetic energy. When a spacecraft is in orbit and starts to fall, additional speed is gained by using the engines to accelerate that speed. The inventor of the reaction engine was a Romanian inventor, Alexandru Ciurcu (1854-1922). Along with a colleague called Just Buisson, they used a reaction engine, tied onto a boat, and sailed up the Seine River on 13 August 1886. The history of rocket engines is so intertwined with spacecraft (astronautics) that it is difficult to separate the two fields. Therefore I will leave this section here.

The second type of engine is the electrical engine. There are three types of classic electrical engines: magnetic, piezoelectric, and electrostatic. Most electric motors operate through the interaction between the motor's magnetic field and electric current in a wire winding to generate force in the form of torque applied on the motor's shaft. The first type of electrical motor that I want to discuss is the magnetic motor. A permanent magnet motor is a type of electric motor that uses permanent magnets in addition to windings on its field, rather than windings only.

Permanent magnet motors rely on neodymium (a chemical element with the symbol Nd, a silvery metal that tarnishes in air and moisture). The neodymium magnet is a mixture of iron, boron, and neodymium. There are different magnets, and the reader can find out more for themselves. The permanent magnet motors mentioned earlier can be used in various forms, such as computer disk drives, hybrid cars, cordless tools, and even some wind turbine electric generators. There is an interesting history regarding hybrid cars. The first hybrid car was developed in 1898 by Dr Ferdinand Porsche (mentioned earlier). However, fuel cars were easier and less expensive to build than hybrid cars. It was not until the 1990s that hybrid cars became popular due to fuel cars' effect on the planet. The hybrid car industry is a multi-billion pound concern, and new models are being produced every year. The next electric motor that I want to discuss is the Piezoelectric motor. This motor makes electricity by applying mechanical stress (pressure or heat) to various elements (piezoelectric material), including ceramics, bone, and crystals. This electrical circuit makes acoustic or ultrasonic vibrations in the material mentioned, which produce linear or rotary motion. The first practical application of this motor was sonar during World War One to detect enemy submarines. There are, in fact, two different types of sonar. Active sonar happens when an acoustic signal or pulse is emitted into the water. When an object is in the pulse's path, an "echo" will be transmitted back to the receiver, called a transducer. The strength of the signal is measured, which will be the time between the signal being emitted and the reception of the signal. The receiver or transducer can then measure the range and orientation of the object. Passive sonar is used mainly to detect noise from marine objects, such as submarines, icebergs, and whales. Unlike active sonar, passive sonar does not emit its signal. This action is an advantage in times of war or when a scientific mission wishes to "listen" to the object. The major disadvantage of passive sonar is that it cannot detect the range of an object unless other passive systems are in use at the same time. This action would allow triangulation of the object to take place. Post World War Two, there was a race to develop better piezoelectric materials, such as quartz crystals, that could be used for radio and television. A Japanese inventor named Isaac Koga (1899-1982) led the field in the development of quartz crystals in the field of communication. The last electric motor that I want to discuss is the Electrostatic motor, which uses the attraction and repulsion of an electric charge to generate energy. Another term for the electrostatic motor is a capacitor motor, most commonly used in micro-mechanical systems to create a drive voltage of fewer than 100 volts. The motors are generally used in air conditioning units, forced air furnaces, and even spa pumps! Air conditioners use refrigeration to cool indoor air using a chemical compound called refrigerants. I have tried to outline below a simple explanation of the system. The cool refrigerant gas is pulled into a compressor, which increases the pressure of the gas, turning the refrigerant into a hot liquid. The heat from this liquid is vented to the outside air. The liquid then goes into an expansion valve, which changes the liquid into a gas. This gas then travels into an evaporator coil that transfers cool air into the building. This gas then travels into a compressor, when the whole system starts again. There are many refrigerants available (the reader can find out more for themselves). However, I would like the reader to be aware of certain chemicals currently being phased out due to climate change. These are known as CFCs (chlorofluorocarbons) and HCFCs (hydrochlorofluorocarbons). We will discuss each in turn. The chemicals known as CFCs (chlorofluorocarbons) are a group of manufactured odourless, non-toxic, and non-flammable chemicals. The elements within the chemicals are a mixture of carbon, fluorine, and chlorine. The chemicals known as HCFCs (hydrochlorofluorocarbons) are similar to CFCs; however, hydrogen has been added to the compound. There have been studies by scholars, including American chemists F. Sherwood Rowland (1927-2012, Nobel Laureate) and Mario J. Molina (born 1943, Nobel Laureate) and Dutch chemist Paul Crutzen (born 1933, Nobel Laureate), which have shown that ozone depletion has occurred due to the release of these chemicals (the reader can find out more for themselves).

The third type of engine I want to discuss is the ion drive, also known as the EM/Cannae drive. Another colloquial name is the Impossible Drive, as the engine works by not using exhaust or propellant. This engine breaks Isaac Newton's Third Law (To each action, there is an equal and opposite reaction).

The E-M drive is short for electromagnetic propulsion drive. Electromagnetic waves are fuel, as microwave photons produce thrust within an enclosed capsule. The inventor of the EM drive is Roger Shawyer, who used to work at Marconi Space Systems (we will be discussing space in a later chapter). The reason why this work is so popular at the moment is due to no weight restrictions. The rocket fuel (explained later) required to power a spacecraft to Mars would require an enormous rocket for the amount of weight. Once a spaceship has broken through the Earth's atmosphere into interstellar orbit, another form of propulsion is required for a long journey. At present, renewable solar energy is the only alternative option. However, the EM-Cannae drive will mean that no other energy source will be required! I have left this section here as there are ongoing tests to see whether the system works.

The author's sincerest wish is that alternative engines and fuels are found during my lifetime.

CHAPTER SIX – THE LETTER F

There was only one material to mention in this chapter, and that was fuel. We will only deal with fossil fuels in this chapter. Any engineer that wishes to know more about the modern world should know the basics of how that world is powered. Two types of fuels are present today, split into chemical and nuclear fuels. Chemical fuels can be broken down into solid fuels, liquid fuels, gaseous fuels, and alcohol. Nuclear fuels are self-explanatory and will be covered in a separate volume. I will now try to explain briefly how some of these work and their advantages and disadvantages.

I will now discuss various solid fuels. The first solid fuel that we will discuss is wood. However, we first have to look back before we can look forward. Three things were required from the first time that man created a small dwelling for the family and livestock. The dwelling had to be waterproof, it had to be near food and water, and the dwelling had to be warm. Fires were lit in the centre, and food was cooked over the fire, which meant both heat and food were provided. This fire is called direct heating; in other words, natural materials such as wood were burnt to create heat around the dwelling. It is now popular in many modern houses to have a wood-burning stove. However, there are a couple of safety issues that should be borne in mind. When a fire is burning in an enclosed room, all the oxygen in a room is exhausted. The effect of not having an opening can be the build-up of carbon monoxide, a colourless, odourless, tasteless gas that can kill when inhaled, as the brain is starved of oxygen. Various scientists throughout the years made the actual discovery of carbon monoxide. Joseph Priestly (1733-1804) is credited with discovering the gas; however, he did not realise that the gas was toxic. In the year 1800, a Scottish chemist called William Cruickshank (died 1810) discovered the toxicity of the gas. The other problem that should be mentioned is that wood-burning stoves can cause the build-up of soot and creosote salts, which, when combined with water from a damp chimney, can lead to sulphuric acid (H_2SO_4). This liquid is a colourless, clear corrosive acid that will eat into the flue lining and even the brickwork of a chimney. Wood is known as biomass fuel; in other words, a fuel that comes from natural sources and is one of the main sources of biomass energy. Wood as a biomass fuel comes from using the waste products of wood, such as shavings, dead trees, tree branches, wood chips, and bark and sawdust. Another wood waste comes from furniture making and wood products used in construction. The old system of using biomass wood was to load the waste wood into big trucks and deliver it to a biomass power plant. Here the biomass is then fed into a furnace where it is burned. The heat is used to boil water in the biomass boiler, and the energy in the steam is used to turn turbines and generators. In the new system, the wood is gathered and transformed into pellets. Wood pellets are formed when the waste product, for example, is put through a crusher called a hammer mill. This action compresses the wood, and it is fed through a press with holes that make up the pellets. The advantage of pellets is that they have a lower moisture content than raw wood; they are uniform in structure and small enough to be transported over long distances. Plant products such as different types of grasses, hemp, corn, and bamboo can also be used as biomass fuel. These can also be pelleted for fuel and added to wood pellets in some cases. These pellets are then burned to create electricity. There are advantages to using wood as a biomass fuel. The main advantage is that the need for fossil fuels is greatly reduced, as the product is a renewable energy source (provided that it is managed correctly). Another advantage is that there are less likely to be severe forest fires, as the by-product of logging, for example, is cleared away, and there is less fuel for the fire to burn and less risk to communities in the area. A forest can get overcrowded with trees, by clearing away trees enables the trees that are left behind to expand into the vacant space that is left behind. There will also be fewer landfill sites if the wood taken is used to produce electricity. The main advantage is that there are fewer greenhouse emissions, as the fuel burned restores carbon dioxide originally in the atmosphere absorbed by the trees. A landowner can use part of their land for growing trees that can be used for biomass energy. The money earned can be partly used for replanting trees so that the system is both renewable and sustainable. However, it may take a long time for the trees to recover the CO2 that the biomass plant has released. The main disadvantage is that a large amount of wood is required to fire a wood-fired biomass plant; therefore, large woodland areas are needed to supply the plant. To harvest an area for wood to be used as fuel in a biomass boiler takes a great deal of heavy machinery, and transportation links have to be in place to move the raw product to the plant; this can be expensive. Wood-fired biomass plants do not produce as much energy as traditional fossil fuel plants, so there are limitations on their use as a sole energy source in an area. The cost of a wood-fired biomass plant can be expensive. Wood has been used as timber for fuel, building houses, furniture, flooring, and hundreds of other uses. Wood, although a natural resource, is not infinite. We need to keep replanting trees. The disastrous effects of deforestation have affected our planet in a cataclysmic way, as there are fewer trees to absorb greenhouse gases. To put it simply, if we do not start planting trees now, then there will be serious consequences for our planet. The next solid fuel that I want to discuss is coal. Coal is a black or brownish-black combustible sedimentary mineral. Coal is primarily carbon; however, hydrogen and sulphur are compounds within the coal. About 300 million years ago, the Earth was covered by swampy forests. Vast coal deposits originate in former wetlands—called coal forests— that covered much of the Earth's tropical land areas during the late Carboniferous and Permian times. Those trees then decayed and were covered with soil, with each layer compressing the next layer. As the layers were successively covered, the material called peat was formed. As thousands of years passed by and the layers hardened, carbon-rich deposits called coal was formed in layers known as seams. The first seam is lignite, the youngest form of coal with high moisture content. The coal seam resembles soil and is known as brown coal. The next layer is called sub-bituminous coal, known as black lignite. This coal has less moisture and is mainly used in power-producing stations. The third layer is bituminous coal, the most common form of coal and the most widely used. The last layer in the seam is called anthracite. This coal is the hardest form of coal and is considered the highest-ranked coal in the world. There are many uses for coal, including generating electricity, producing steel, and domestic use. Other coal industries include the cement industry, chemical, and pharmacy industries. The advantage of coal is that it is a relatively cheap fuel, and it is reckoned that coal reserves around the planet will last for another three hundred years. The coal mining industry employs thousands of workers worldwide and is considered a viable industry. The last major advantage of coal is that the reliance on oil from foreign countries is reduced if we use our domestic coal reserves. This action was shown in the 1970s when there was a global oil crisis due to the increase in the price of a barrel of oil. This action was due to the oil-producing countries, such as Saudi Arabia, stopping oil from being exported due to American foreign policy (the reader can find out more for themselves). This led to a worldwide stock market crash, high inflation, and mass unemployment. There was simply no fuel for cars! However, there are massive disadvantages to using coal. The waste produced from burning coal contains carbon dioxide, arsenic, nitrogen oxide, and sulphur dioxide, amongst other chemicals. I would like to draw your attention to the Great Smog of London in 1952. Poor quality coal (lignite) was being burnt as the main fuel source. This smog was due to anthracite being sold to foreign countries to pay off the debts during World War Two. A severe cold snap meant that more coal was being burnt than usual during this time. There were dry, clear skies trapped under warm air (called an anticyclone). This weather meant that fog could form, and with such high pollutants, this turned into smog containing sulphuric acid. In modern terms, we can now measure Particulate Matter (PM), small particles not visible to the naked eye. Although there was no measurement at the time of the Great Smog, official records state that at least six thousand people died due to smog, with current estimates as high as twelve thousand. Another disadvantage is the effect that coal mining has on the environment, with wildlife habitats being destroyed as coal is dug from the ground. The last adverse effect is on the miners themselves. A lung disease known as black lung disease (coal workers pneumoconiosis CWP) occurs when coal dust is inhaled. Continued exposure over a long period causes scarring in the lungs, known as fibrosis, which stops the ability of the miner to breathe. Before leaving this section, I must mention

Aberfan in Glamorgan, South Wales. Every generation of schoolchildren should be made aware of this manufactured disaster that could have been avoided. On 21 October 1966, one of the colliery spoil tips from the local mine collapsed, engulfing the town. This landslide killed 116 (one hundred and sixteen) children and nine adults. The National Coal Board was found to be guilty. I want to insert a structure featured on the "Abandoned Engineering" programme. This place is Porto Flavia, a sea harbour located in Sardinia, Italy. I have mentioned this place as coal (and other materials) were dug from the rock face. Traditionally, the coal was placed into wicker baskets and lowered down into the waiting ships below (there was no safe anchorage, the ships would station themselves at the bottom of the rock face). Once the loading was completed, up to thirty tons in weight, the ship would sail off to a safe harbour where it could be properly unloaded. In the author's opinion, this unloading made the Porto Flavia one of the most unusual coal mines in the world (the reader can find out more for themselves).

I want to discuss the following solid fuel: peat mentioned above. This material results from years of decaying organic matter such as trees and other vegetation found in peat bogs. Therefore, peat can be described as a fossil fuel. Peat has been used as a fuel source for thousands of years, mainly in Finland, Ireland, and Scotland. There are several advantages to using peat as fuel, as it produces as much heat as burning coal and is cheaper than other fossil fuels. Peat has traditionally been cut by hand and stored in blocks to dry. However, there are disadvantages. Cutting peat destroys natural habitats and may even disturb natural breeding grounds. The other point to note is that peat is organic, and therefore the carbon dioxide present in the material can be released into the atmosphere. This action can lead to an increase in carbon emissions, leading to climate change.

In this section, we will be considering liquid fuels. I have split the fuels into petroleum, coal tar, diesel, and kerosene. I will consider a brief history before considering the advantages and disadvantages of each type of fuel. Petroleum is an organic, yellowish-black liquid found beneath the Earth's surface. It is commonly refined into various types of fuels. Petroleum came to prominence principally due to the invention of the internal combustion engine and the rise in commercial aviation. Other uses for petroleum include plastics, fertilisers, solvents, adhesives, and pesticides. However, I would like to include a brief history for students who do not have access to a computer. Petroleum has been in existence for thousands of years. Various countries used it as a wall covering using tar and asphalt. This action can be dated as far back as 4000 years ago in the Kingdom of Babylon. I want to introduce Edwin Drake (1819-80), Colonel Drake. Edwin Drake was an American businessman and the first American to drill for oil. The oil well was situated in Titusville, Pennsylvania, United States. The company that employed Drake was known as Seneca Oil (now bankrupt), and the well was the first oil well drilled in the world. I have included this oil well, as the methods used to extract the oil have been copied worldwide. This place was also the first oil well to use a steam engine (see earlier), and the fact that this well started the oil boom that has lasted to this day. I want to include a few other people here as they play a major part in world events. This next piece comes from the seminal programme "War Factories" on the Yesterday Channel. The Nobel family is famous for Alfred Nobel (1833-96), the Swedish chemist and engineer who famously invented dynamite. The Nobel Prize is named after him. His brother Robert Nobel (1829-96) was a Swedish businessman who was a pioneer in the Russian oil industry in Baku, Azerbaijan. However, it was another brother, Ludvig Nobel (1831-88), a Swedish engineer who managed to make the oil industry a success (inventing the first oil tanker, which will be discussed later). All three brothers were involved in the early days of the Russian oil industry through their company called Branobel (The Petroleum Production Company Nobel Brothers). Before we continue, we must set the story in context. Up to the middle of World War One, all major battles had been fought on horseback, with cannons and swords providing the weapons required (we will discuss munitions in a later chapter). However, with the advent of the internal combustion engine, oil became a prized commodity. In 1912, the Branobel company produced up to seventy-five million barrels of oil, which was half the world's oil supply. However, in 1920, the Russian Revolution occurred (Azerbaijan was originally part of the Soviet Union), and the brothers lost control of the company (there is a fascinating story regarding this company, and I will let the reader find out more for themselves). I have mentioned this area of Azerbaijan as this became a key focal point during the Nazi invasion of Russia in 1941, known as Operation Barbarossa. The reason behind the invasion was the oil fields of Baku, required for Hitler's ongoing campaign of world domination. However, the German army was halted at the Battle of Stalingrad, and oil fields were saved. That was one of the most important battles of World War Two (I will discuss synthetic oil later in the book). The next liquid fuel that I want to discuss is coal tar. This fuel is a thick black or dark brown liquid produced as a by-product of coke and coal gas. It is estimated that over a thousand chemicals form coal tar, so the reader will have to find out for themselves! However, benzene, naphthalene, phenols, and other organic chemicals are all present. Biphenyl (E230) is a naturally occurring compound ($C12H10$) in coal tar, crude oil, and natural gas (all fossil fuels). It is produced as a by-product of benzene and is used as a preservative for citrus fruits to prevent fungus and mould growth. The main uses of coal tar are medicinal for skin conditions (in small doses) and creosote, used to preserve timber as a preservative and a fungicide. Coal tar is also used to fire boilers and in asphalt for roads. There is a famous case in England that I would like to draw the reader's attention to, as the term "caveat emptor" should always be borne in mind. Bawtry Gasworks is a groundwater contamination site in Yorkshire, Northern England. The case is mentioned due to the problems of who would pay for the pollution that happened. The Bawtry and District Gas Company owned Bawtry Gasworks till 1931. The production of town gas stopped in 1952 at the site. To create town gas, coal was burned in an oxygen-starved environment, which as a by-product produced heavy tars contaminated with heavy metals. The pits containing the tars and underground storage tanks were never removed from the site. Before being sold to housebuilders, the site was eventually used as a depot by various owners, including the East Midlands Gas Board. A housing estate was built on the site; however, an issue was found when a resident dug in their garden in 2001 and uncovered a tar pit. The tar pit itself showed that coal tar was present on the site, which is toxic to humans. When this happens, the site may be labelled as a 'special site' so that the responsibility for the clean-up passes from the local authority to the Environment Agency. They will then track down the site owners and order a clean-up operation. However, this is where the problem lies. Bawtry and District Gas Company went into liquidation; the East Midlands Gas Board went the same way when the industry was privatised by the Gas Bill in 1948. The site came under the jurisdiction of British Gas and Transco, but they had never occupied the site. Therefore there was no one to assume responsibility for the clean-up cost of the site. As part of the Environment Act 1990, the blame fell to the housebuilders who should have known about the contamination. However, the housebuilders had also gone into liquidation. Therefore there was no party responsible for the groundwater contamination that had occurred. The cost of the clean-up operation was £695,872, which worked out at £63,253 per householder. The Environment Agency decided it would be unfair to make the residents pay and therefore carried out the work at public expense. All underground structures were removed, and the soil was cut down by 600mm. A separator was installed so that the residents would have no further contact with the contamination. This case shows that care must be taken when developing previously used industrial sites (brownfield land). The next liquid fuel that should be mentioned is diesel. This fuel is named after the inventor, engineer, and entrepreneur Rudolf Diesel (1858-1913). Rudolf Diesel published an essay in 1893 entitled "Theory and Construction of a Rational Heat-engine to Replace the Steam Engine and the Combustion Engines Known Today".

Four years later, in 1897, Rudolf Diesel built the first Diesel engine (modified from his original patent DRP 67 207). The original use of the diesel engine was for factories and small scale industries, as steam engines were costly and not as efficient as his engine. It was only later that the diesel engine was adapted for automobiles. I will now give a very brief history of the diesel engine, as the reader should investigate for themselves. The German company MAN (founded in 1758) in 1910 started production of the first two-stroke engines. In 1912, the first train locomotives with diesel engines were run on the Winterthur-Romanshorn railway line in Switzerland. In 1933, the Junkers company manufactured the Jumo 205, one of the most successful aviation diesel engines. The airship LZ 129 Hindenburg, a Zeppelin airship, used diesel engines for flight. The airship tragically crashed in 1937, losing thirty-six lives in the process. The Hindenburg Disaster was a photograph taken in 1937 by Sam Shere and ended the era of the large airship. The first successful diesel train in America was the Zephyr Pioneer, used between Denver and Chicago. Diesel was the natural choice for American locomotives, as steam engines required too much maintenance and were expensive to run. A person that also deserves mention is Clessie Cummins (1888-1968), a businessman and entrepreneur who improved diesel engine design. The Cummins engine set five world records for endurance and speed. In 2006, the four-wheeled drive diesel engine was the JCB Dieselmax, a British racing car that holds the diesel engine land speed record at over 350 mph. I will now discuss the advantages and disadvantages of diesel engines. The difference between a normal gasoline engine and a diesel engine is as follows. A normal gasoline engine requires a spark (combustion), whereas a diesel engine works on hot air compression to ignite the fuel inside. This action means the engine gives better fuel economy and requires less maintenance. The other advantage of diesel fuel is that diesel engines tend to last longer and can pull heavier loads than a conventional engine. However, there are disadvantages as well. Diesel engines do not like cold weather, as diesel fuel starts to gel below minus twenty degrees Celsius. However, the worst disadvantage is the emissions from diesel engines, as these can contain such elements as nitrogen oxide and dioxide, sulphur dioxide, and other harmful by-products. These can all lead to air pollution. I want to discuss that the next liquid fuel is kerosene, also known as paraffin and lamp oil. This fuel is pale and yellow (can also be colourless) liquid derived from petroleum. Kerosene is a combustible organic fuel. Its name derives from Greek: κηρός (keros), meaning "wax". This fuel was registered as a trademark by Canadian geologist and inventor Abraham Gesner (1797-1864), a Canadian physician and geologist who invented kerosene in 1853. The term kerosene is common in Argentina, Australia, Canada, India, New Zealand, Nigeria, and the United States. In contrast, paraffin (or a closely related variant) is used in Chile, eastern Africa, South Africa, Norway, and the United Kingdom. The term lamp oil, or the equivalent in the local languages, is common in most of Asia. The most common usage of kerosene is in the aviation industry to power jet engines and some rocket engines (discussed in another chapter). Kerosene is also used all over the world as cooking and lighting fuel. The British Standard BS 2869:2017 covers the fuel used in agricultural, domestic, and industrial applications. I will now cover a brief history of kerosene. A mention should be made at this point of James Young (1811-83), a Scottish chemist best known for his method of distilling paraffin from coal and oil shale. Chemist James Young noticed a natural petroleum seepage in the Riddings colliery at Alfreton, Derbyshire, from which he distilled a light, thin oil suitable for use as lamp oil while obtaining a more viscous oil suitable for lubricating machinery. The production of these oils and solid paraffin wax from coal formed the subject of his patent dated 17 October 1850. His achievements also included the discovery of how to distil kerosene from seep oil, the invention of the modern kerosene lamp (1853), the introduction of the first modern street lamp in Europe (1853), and the construction of the world's first modern oil well (1854). When oil was discovered in Pennsylvania, the United States, in 1859, there was a large amount of excitement around the discovery. This discovery led to the oil booms, which started the oil craze, forming modern life. Kerosene was used in the early 1900s for lighting and heating in domestic houses and small scale industries. However, the onset of electricity and petroleum meant that kerosene was not used for larger vehicles. There are several advantages to using kerosene as a fuel (according to the website, "It still runs"). Kerosene is non-corrosive; therefore, it can be stored in plastic containers for up to a year and even longer in metal containers. Kerosene does produce fumes, though less than other alternative fossil fuels, so a well-ventilated area is advisable. Kerosene is easy to ignite; all required is a spark, which makes its use as a fuel better in countries with poor electricity grid networks. Kerosene has another major advantage; the fuel is cheap to purchase and is available in small quantities. The last major advantage was that kerosene, made first from coal and oil shale, then from petroleum, has taken over a lucrative lamp oil market. As kerosene production increased, whaling for fuel declined (discussed in the next section). Some disadvantages have been mentioned already. Kerosene produces fumes, including nitrogen oxide, sulphur dioxide, and carbon monoxide. These fumes are carcinogenic, which means they are harmful and even fatal to humans. I have mentioned that kerosene is highly flammable and explosive; therefore, great care must be taken to ensure no ignition point near kerosene.

This section will concentrate on those fuels derived from gas, including LPG, naphtha, Natural Gas, and coal gas. This gas is by far the biggest group of gas that we have discussed in this chapter. The person behind this invention was an American chemist called Dr Walter Snelling (1880-1965). I will discuss each of these before discussing the advantages and disadvantages of the material. LPG is short for Liquefied Petroleum Gas. This gas is a hydrocarbon (mixture of hydrogen and carbon) gas that exists in a liquid form (although this should be in the previous chapter, it is a gas). LPG is used as a fuel in cooking, industry, vehicles, refrigerant, and as a propellant (a material used for the production of energy). LPG is also known as butane, propane, propylene, butylene, and isobutylene. I will now discuss each of these briefly. Butane is a gas at room temperature and was discovered in 1849 by the British chemist Edmund Frankland (1825-99, who also discovered helium). When butane is mixed with propane, it is referred to as LPG. Another use for butane is a refrigerant for air conditioning systems (a mixture of isobutane, a pure form of butane and propane). This gas is preferred to HCFC (Hyrdocarbonfluorocarbons), which affect the ozone layer. Butane is also used for outdoor cooking. Propane is a gas at room temperature formed as a by-product during natural gas processing. Propane was discovered by the French chemist Marcellin Berthelot (1827-1907). His theories on thermochemistry underpin our modern world today. The main uses for propane are outdoor cooking and fuel in buses and taxis. Propane can be transported easily in containers, therefore is used as fuel in rural areas with no mains gas supply. Other uses for propane are flammable gas in blowtorches, hot air balloons, and propellant in aerosol sprays. Another form of LPG is propylene, also known as propane. This gas is formed during the steam cracking of large hydrocarbon molecules (such as crude oil). The main use of propane is in the manufacture of polypropylene, a plastic that is widely used today in containers, carpets, and other materials. The final form of LPG that I would like to discuss is butylene and isobutylene (a pure form of butylene). Butylene is also known as butene and is formed during the cracking process of crude oil. The main use of butene is in the formation of synthetic rubber. As an aside, synthetic rubber eventually replaced natural rubber in the production of car tires. There is an abandoned ghost town in Brazil called Fordlandia. Natural rubber trees were planted, and a whole town was built for the workers of the rubber plant. The whole project is estimated to have cost the Ford company over $20 million. The project was a disaster, native pests ate the trees, and the advent of synthetic rubber meant the project was abandoned. Isobutylene is used to produce Methyl tert-butyl ether (MTBE), a fuel additive to vehicles to reduce engine knocking and reduce harmful emissions. The next gas fuel that I want to discuss is naphtha. This material is formed from natural gas concentrates and distilled from petroleum. The uses of naphtha are varied, and one of the most useful is the distillation of heavy crude oil. Undiluted crude oil is difficult to transport through pipelines and tankers due to the thick viscosity of the liquid. Therefore, naphtha is added to the mixture and is extracted during the refining process. Other uses for naphtha are plastic production, Zippo lighters, and a solvent for paint. However, the disadvantages of naphtha relate to the volatile and flammable nature of the product. How volatile a product is, relates to the speed at which the gas vaporises at room temperature. The higher the volatile nature of the gas, the probability that the gas will ignite will increase. Naphtha is also highly flammable; therefore, great care should be taken in storing the gas. Another point to note is that naphtha should never be inhaled, as this can be fatal. The next fuel that I want to discuss is natural gas. This gas is composed mainly of methane and is formed during the breakdown of vegetation in the soil. Therefore, natural gas is a fossil fuel and has been used for thousands of years. Natural gas was seen as an unwanted by-product of oil drilling and was generally burnt off or disposed of by the oil company. Natural gas is also found in shale deposits, and the process of removing the gas is called fracking. However, due to the chemicals used and the nature of removal, this remains a highly contentious issue. The modern uses of natural gas are cooking, heating, electricity generation, and fertiliser production. Natural gas, in its natural state, is odourless, colourless, and tasteless. Due to the odourless nature of natural gas, an odoriser is added called mercaptan. This gas smells like rotten eggs, so leaks are easier to detect. It should be noted that Hydrogen Sulphide (H2S) gas is also found during the drilling for oil and gas. However, this chemical is corrosive to pipelines and poisonous to humans, even fatal in high concentrations. Natural gas is turned into a liquid (LNG) at a liquefaction plant (the reader may want to find out more for themselves). The material LNG can now be transported through pipelines and trucks to where the material is required. The other form of natural gas is compressed natural gas (CNG), which is used mainly in the automobile industry. However, it should be borne in mind that natural gas has a high environmental impact, as methane is a greenhouse gas that contributes to global warming. Natural gas is also volatile, and there are incidents of explosions every year due to the undetected build-up of gas, and an ignition source nearby. There is one last major disadvantage, with fatal consequences. Natural gas heating systems contain carbon monoxide, which is also odourless, colourless, and tasteless. Carbon monoxide poisoning is usually fatal; therefore, a well-ventilated space is recommended. The next fuel that I want to discuss is coal gas. Gaslighting can mainly be attributed to the Scottish engineer William Murdoch (1754-1839). However, his work has been overshadowed by the works carried out by his employer, Boulton and Watt (who specialised in steam engines). The first commercial gas works were the London and Westminster Gas Light and Coke company (1812-1949) in Horseferry Road, Westminster (British Gas came from this founder company). The company laid wooden pipes to illuminate Westminster Bridge in the year 1813. The next big step in the history of coal gas was the illumination of factories, replacing the outdated oil lamps and candles. This gas made night shift work possible, which rapidly increased production levels. In 1926, the Gas Light and Coke Company opened a training college for apprentices in Battersea, south-west London. This course enabled the industry to expand and teach safe methods. The uses of gas in the modern world are primarily cooking and heating. The evolution of electricity meant that gas-lit lamps soon became outdated. The by-products of coal gas are coke, coal tar (used in highways), benzole (a type of motor fuel), and creosote, a well-known wood preserver. However, there are also unwanted by-products from producing coal gas. Coal tar is a by-product, and there is a ground contamination site that I have mentioned previously. The problem with old gasworks sites is that the ground is heavily contaminated after the activity has ceased. The clean-up operation is costly and time-consuming, making this fuel product almost obsolete. I hope this chapter helps the reader understand gas fuel. It is a large and complex process, and I would advise the reader to undertake further research should they wish to.

I hope this chapter on fossil fuels sheds some light on why this type of fuel was considered so crucial up to the end of the twentieth century. It has now come to the world's attention that these fuels are finite, polluting materials. The following energy sources we have to use should be renewable, sustainable, and non-polluting if we are to preserve the planet.

CHAPTER SEVEN – THE LETTER G

This chapter considers the letter G. The first thought that came to the author's mind was gunpowder, which contains saltpetre, charcoal, and sulphur. As an aside, the BBC Series called "Taboo", which aired in 2017 contains a plot that involves making gunpowder. This material has been researched, and apparently, the formula given works! However, there are many inherent dangers when making gunpowder, so please be aware. Gunpowder belongs to a group of substances called explosives. These substances react with a large amount of energy when suddenly released. The potential energy is released in heat, light, and sound. However, some materials are explosive when subjected to pressure or merely ignite in certain circumstances. There are two types of explosives, called low explosives and high explosives. The low explosives group, to which gunpowder belongs, burns at subsonic speeds. High explosives, such as TNT and dynamite, detonate at supersonic speeds sending a shock wave through the air. A person that should be mentioned is Dr James Gale, a blind Victorian physician who lived in Devon. Dr Gale was an inventor who proposed a safe way of storing gunpowder. His theory of mixing gunpowder with powdered glass to render it non-explosive seems plausible, though people never acted on his ideas. This chapter will only include munitions such as cannons and firearms. For those students who do not have access to a computer, I will now give a brief history of gunpowder. This history will be followed by some historical examples of the devastating power of gunpowder. Gunpowder was invented in China around the ninth century AD. As China opened for trade, using the Old

Silk Road, gunpowder came to Europe and other parts of the world during the thirteenth century. The reader should remember that any army with access to gunpowder had an immediate advantage over their enemy forces. Throughout medieval times, wars have been fought on horseback, and the sword was the only weapon that could be used. The enormous army, though not always, won the battle. History is full of battles that have been won, either by superior numbers or by using the natural terrain to their advantage. It was gunpowder that changed the face of war forever. There is a more affable side to gunpowder, apart from war. Fireworks have gunpowder, and the colours seen on a firework display, especially at Bonfire Night (see later), are gunpowder mixed with various chemicals to produce the colours we all enjoy.

As an aside, there is a gunpowder mill located at Chilworth, near Guildford, Surrey. There is an exciting history of the mill, which the reader can find out for themselves. The mill was established in 1626 by the East India Company (the reader can find out more for themselves). The mill supplied gunpowder to government forces during the English Civil War in 1642 (the reader can find out more for themselves). In 1882, the smokeless (brown) powder was manufactured on-site. In 1901, there was an explosion at the granulating (corning) part of the works, in which six men were killed. The mill closed in 1920, ceasing all gunpowder manufacture. The other gunpowder factory I would like to discuss is Waltham Abbey Royal Gunpowder Mills. This gunpowder factory started production around the end of the seventeenth century and was in use for over three hundred years, ending production in 1945. A high point in the site's history was in 1787 when the Crown purchased the site. The French Revolution and the Napoleonic wars (the reader can find out for themselves) saw a massive rise in demand for gunpowder. Both World Wars saw production increase till new technology made gunpowder obsolete.

I will now relate some historical information that I hope the reader will find interesting. Castles were often under siege as enemies sought to take over the Castle during medieval times. However, the solid fortifications and thick walls were an immediate defence against any army. Therefore, castles were considered the defence of choice during this period. The foundations of these castles were dug down to bedrock, where the walls would be built directly off this type of soil. If there was no bedrock or the land was too poor, a foundation trench would be dug, filled with rubble, and built the walls. Therefore it can be seen that foundations played an essential part in early medieval construction. If there were no hills in the area, the earth was piled onto land to create an artificial hill called a motte. A bailey was a levelled courtyard located next to the motte. On top of the motte usually sat a keep (a fortified tower). The motte and bailey were usually protected by a ditch encircled the Castle. This ditch was usually filled with water and was called a moat. The Castle was protected by wooden stakes that surrounded the bailey. This Castle is called a motte and bailey castle and was one of the original designs for castles. These have been altered over time, and stone was used for wall construction. They can be seen throughout England and Europe. However, these defences would prove inadequate against the cannon. I want to mention Corfe Castle, a scheduled Ancient Monument. Corfe Castle is situated north of the village of Corfe in Dorset and is an example of a motte and bailey castle. A deep ditch separates the town from the Castle, which may be why the site was initially chosen. The Castle is built of Purbeck stone, ashlar (a finely cut piece of masonry that has been cut square), and rubble. There is evidence of flint in the core of the walls. As with most castles, the Castle grew in size over successive years. There is an inner ward, west bailey, and outer bailey. The original part of the Castle can be seen in buildings around the keep that were indeed defensive walls and a building in the middle pre-1100. The English Civil War (1642-1651) was between the Cavaliers (or Royalists), who supported King Charles I, and the Roundheads, who supported the Parliamentary forces under Oliver Cromwell. In 1645 the Castle came under siege from Parliamentary forces, and the Castle was captured. In March 1646, a vote was passed by the House of Commons for its demolition. Explosives such as gunpowder were laid at the base of the walls, and the Castle has been a ruin ever since. It has been maintained by the National Trust ever since. The most well-known use of gunpowder was the Gunpowder Plot of 1605. This plot was a plot by English Catholics, led by Robert Catesby, to assassinate King James I of England. The idea was to blow up the House of Lords during the state opening of Parliament when the King would be present. A letter was sent to William Parker (1575-1622, 13th Baron Morley, 4th Baron Monteagle), informing him of the plot against the King. On 4 November 1605, the cellars underneath Parliament were searched.

The conspirator Guido (Guy) Fawkes was discovered with thirty-six barrels of gunpowder. All of the conspirators found were hanged, drawn, and quartered. The celebration of the thwarting of the plot is known throughout England as Bonfire Night. There have been some famous churches that have been destroyed after being hit by lightning. On 27 October 1697 in Athlone, Ireland, the Castle was struck by lightning. It had been storing 260 barrels of gunpowder when lightning struck. Most of the Castle was destroyed, and in surrounding buildings, however, only eight people lost their lives. In 1769, the spire on top of the Church of the Nazaire in Brescia, Italy, was struck by lightning. The church was storing 100 tons of gunpowder in its vaults for safekeeping. The lightning travelling through the building ignited the gunpowder. The explosion destroyed parts of the city, and three thousand people were killed.

I apologise that this is such a brief chapter; however, I hope the reader enjoyed this.

CHAPTER EIGHT – THE LETTER H

There is an option that I would like to discuss, an option that fits with the letter H. The option is to heat. I have mentioned the thermal engine in previous chapters, so I will not repeat the details here.

Ever since ancient times, heat has changed the world we live in beyond compare. I will not try and go through the old Stone Age to Bronze Age times, as this would take up too much space in this book. We will consider the history of heat, as this will lead to some exciting places! I want to consider in this chapter how heat has influenced the built environment. This is an evolution process, how natural materials were eventually superseded due to their width and height limitations (please see below). Eventually, manufactured materials such as brick and steel (mentioned in another chapter) became prominent and changed the world.

Therefore, we need to understand the first natural building material, stone. There are, in fact, three types of rock: Clastic (rocks that are composed of broken pieces of older rock), chemical (formed when dissolved materials fall due to a chemical solution) and organic (formed from plant or animal debris). In the interest of engineering, I would like to discuss some structures made out of stone. The tallest building in ancient times was the Great Pyramid of Giza, Egypt, which was 146m high and was built in the 26th century BC in stone with a white limestone top layer. The Great Pyramid was created using granite stones, and there are over two and a half million stones in the Pyramid. There was white limestone covering the Pyramid, which would have made the Pyramid an incredible sight. It is one of the Seven Wonders of the Ancient World. However, the Great Pyramid of Cholula in Mexico, built by the Mayans, is the largest in the world. It is 66m high; however, the structure's base measures 450 feet by 450 feet, four times the Great Pyramid's size. The Cholula Pyramid has a total volume of 4.45 million cubic metres, which is twice the size of the Great Pyramid of Giza. As far back as ancient times, the bottom half of a building was given a broader base to counter the seismic waves of an earthquake. There was a direct correlation between the width and length of a Pyramid and the Pyramid's height. The Pyramid can be the safest structure, as it has a broad base compared to its height. The other main drawback is that a pyramid cannot extend outward; there is a confined area once the base is laid. The Pantheon in Rome is the following example of using stone as a material. It is one of the finest Ancient Rome buildings still standing in modern Rome, dating from ad 118. The building comprises an entrance, called a portico, with columns supporting the weight. Behind the entrance is the Rotunda, a circular dome supported by columns. The dome was at one time the world's largest unreinforced concrete dome. The dome is 43 metres high and spans over a space that is 43 metres in diameter, making it the largest dome in the world until modern times. It has a circular opening at the top, called an oculus, eight metres across. The concrete at the bottom is denser than at the top using different aggregates (stones). The bottom of the dome is heavier by using granite stones with brickwork providing a counter-balance. The top of the dome is lighter by using lightweight volcanic stones, called pumice, and that is how the dome is constructed. The castle was seen as the best fortification during medieval times to ward off enemy attacks. The best example is Beaumaris Castle, located on the Isle of Anglesey, off the coast of Wales. It is a concentric castle, made up of circles of stone defensive walls, with a central keep in the middle. The inner walls were higher than the outer walls, which enabled archers and other defensive measures to be used. Concentric castles have battlements, a drawbridge, a vigorously defended entrance and arrow slits. Beaumaris Castle s considered one of the finest examples of 13th and 14th-century military-style architecture and is a UNESCO World Heritage Site. It is the last and most significant stone concentric castle built by King Edward I (called Longshanks, due to his height, who reigned between 1272 and

1307). This castle is perfect in geometric terms, which means that the castle was built to exact measurements. The castle's building probably started around 1295 to ward off the threat posed by the Welsh in the region. Edward I died in 1307, and building work on the castle stopped in 1330. Building the castle was astronomical, around £15,000 (roughly £80 million in today's money). I hope these examples show the reader how stone was used up to medieval times. The modern world requires new products forged by heat.

I want to fast forward a few centuries and discuss a building material that changed the world. As the world became populated, more buildings were required to house that population. New material was used in more significant quantities than before, and that material was brick, made from clay. I will not bore the reader with a history of how bricks are made; there is plenty of information on a computer near you! Instead, I believe the best way to learn is to look at practical engineering examples. The first example is the Shwedagon Paya pagoda, located in Mynamar (formerly Burma). The pagoda is located on top of a hill and is ninety-nine metres (325 ft) high. This pagoda is a building rumoured to be over 2,600 years old. The structure is made entirely of bricks, then plastered over and later gilded with gold. Pagodas are incredibly interesting for engineers. The actual structure of a traditional wooden pagoda is made up of layers, a tower of boxes with joints that slide together (like a snake dancing); these are called mortise joints. The other secret to a pagoda's success is inserting a column at the centre that will stop the storeys from falling over. Another famous Wonder of the Modern World that has bricks in its construction is the Great Wall of China. The Great Wall that we know today was built during the Ming dynasty (1368-1664 AD), and new materials of sun hardened clay, called brick, were used. One surprising ingredient used in the mortar mix was rice due to the sticky nature of the rice used. The Ming Dynasty part of the Wall had a height ranging from five to eight metres (sixteen to twenty-six feet); the width of the Wall was about six metres (twenty feet) at the bottom to five metres (sixteen feet) wide at the top. There were over twenty-five thousand watchtowers along the Wall. The last non-domestic example I want to discuss is the brick-built shot tower. There are examples of this type of tower in Europe and the United States. These buildings were designed for the manufacture of lead balls that were used in muskets. Molten lead was poured down from the top through a channel and naturally cooled as it reached the bottom of the tower. The lead was collected in a water-filled tank at the bottom of the tower, and the balls were quality checked for their round shape. The Phoenix Shot Tower in Baltimore, United States, was featured in the "Abandoned Engineering" programme and is worth investigating. Concerning domestic architecture, brick was a common material for all houses in the eighteenth century, known as the Georgian period. This period was a period of global trade, and housing became a method of showing wealth. This period was the age of the terraced townhouse, and semi-detached houses became prevalent. A great example of Georgian architecture is the crescent of houses in Bath, known as The Circus.

The last building material that I want to discuss that can change the face of a building is glass. Glass can transmit, reflect and refract light (refraction, by definition, is when sound or light waves pass through a substance, usually glass, speed up and change direction). In order to discuss the evolution of glass (I will not go into a long history), firstly, we need to mention stained glass windows. In the book Pythagoras in the Corner Introduction, I mentioned stained glass windows. There are buildings, including cathedrals like York Minster, with unique stained glass windows. The most extensive stained-glass map in the world is located at the Mary Baker Eddy Library in Boston. The globe has 608 stained glass panels and is three storeys high. It has a thirty-foot long bridge running through the centre, giving a three-dimensional feel to the structure. However, the glass that I want to discuss is another type of glass, called architectural glass. This type of glass is also known as float glass. This engineering process involves floating molten glass on a bed of molten metal. The glass is fed through a kiln before being cut by machines. This glass technique was developed in the 1950s by a firm called Pilkington, based in Merseyside. This glass is the most widely used globally, both for commercial and domestic use. The Louvre Museum in Paris has a glass and metal pyramid entrance to the building. This Pyramid was commissioned in 1984 and was completed in 1989. There are 673 panes of glass in a structure that measures over twenty-one metres (71 ft) high and has a base of 35 metres (115 feet) across. The design was due to the difficulties in using the original entrance to the Museum. There are two other pyramids around the Museum, including an inverted pyramid that points down into the Museum shopping arcade. The

Gherkin (30 St Mary Axe), or the Swiss Re building in London, to give it its proper name, is a skyscraper that is curved in nature, using glass and steel in construction. There are 35 kilometres of steel in the building with over 38 floors to climb. The building took two years to complete and cost 204 million pounds. It is 180 metres in height and has 7,429 panes of glass. The other building that uses glass to significant effect is the IAC Building, Manhattan, New York. This building is the headquarters of the InterActiveCorp Group (a media company), and the building was completed in 2007. The building is ten storeys high; however, the unique design of a larger bottom half resembling a beehive and the top half smaller does not seem to be the case. There are enormous thirty-five feet by twenty-two feet glass panels, and the windows in the building are full height, reaching from floor to ceiling and fading from clear to white, adding to the overall design. It is when darkness falls that the building comes to life. The easy, fluid exterior, like full-blown sails, is a marked contrast to the traditional skyscrapers of New York.

I hope this chapter helps the reader understand the importance of heat. I have specifically concentrated on the evolution of building materials. I believe that it is essential for every engineer to understand the materials they work with to create their unique structures. Engineers do not work in isolation, and neither should their buildings. Every building, in theory, should live in harmony with the existing infrastructure and built environment. I do not mean that all buildings should look the same, as that creates a bland environment. Instead, buildings should complement their surroundings (Lesson over).

CHAPTER NINE – THE LETTER I

There were plenty of options for this letter; however, the choice is iron. The reason is that iron is the 4th most abundant element in the Earth's crust. Iron is the sixth most abundant element in the Universe. Experts have accepted the Earth's core to comprise primarily of iron. Earth's solid inner and liquid outer cores are primarily composed of iron (approximately 85 per cent and 80 per cent by weight, respectively). According to the Internet, the electric current generated by the liquid iron creates the magnetic field protecting Earth. Iron is found in the cores of all of the planets in the Solar System. According to the Internet, iron is the heaviest element formed in the cores of stars. Elements heavier than iron can only be created when high mass stars explode (supernovae). Iron derives its name from the Anglo-Saxon' iren'. According to the Internet, iron may be derived from earlier words meaning "holy metal" because iron was used to make the swords used in the Crusades. The element symbol, Fe, was shortened from the Latin word 'ferrum', meaning 'firmness'. Iron is found in meteorites, and because these fall from the sky, some linguists have conjectured that the English word iron (Old English īsern), which has cognates in many northern and western European languages, derives from the Etruscan aisar, which means "the gods". Some have linked the iron in meteorites to a verse in the Quran(57:25) that says, "… and We sent down iron in which is incredible strength and many benefits for mankind." There are two different types of iron that I would like to discuss below. An iron alloy is steel, which I will discuss in a later chapter.

I want to give a very brief history. The first signs of the use of iron came from artefacts of the Sumerians and Egyptians dated to around 4000 BC. They prepared tips of spears, daggers, and ornaments from iron recovered from meteorites. Iron objects were found in Egypt around 3500 BC. They contain about 7.5% nickel, which indicates that they were of meteoric origin. Turkey's ancient Hittites of Asia Minor were the first to smelt iron from its ores around 1500 BC, and this new, more robust metal gave them economic and political power. Around 2000 BC, the population of Europe moved from smelting copper to harnessing iron, and the transition from the Bronze Age to the Iron Age began. Christopher Wren used iron hangers to suspend floor beams at Hampton Court Palace and iron rods to repair Salisbury Cathedral and strengthen the dome of St Paul's Cathedral. I will leave this section here.

The first type of iron is wrought iron, which is tough, malleable, corrosion-resistant and easily welded. The iron is moulded whilst hot from a furnace and therefore was called wrought iron. A name that should be mentioned is Abraham Darby (1678-1717), who was the first person to develop a method of producing pig iron in a furnace. This material was an essential step in the production of iron for the Industrial Revolution. Henry Cort (1740-1800) should be mentioned here, as this man was one of the first people to refine iron. The base product of pig iron was converted to wrought iron by using a furnace. The "puddling" process, which Henry Cort refined, stirs iron when hot in a bath of liquid open to the air. The iron is extracted and rolled into bar iron, ready for use. The system was limited to producing small amounts and, accordingly, was superseded. However, wrought iron is used to produce many products, from fencing and balconies to gates and windows. I want to take this opportunity to consider some famous examples of wrought iron products. The most famous example of wrought iron has to be the Eiffel Tower in Paris, France. The structure has over two and a half million rivets that hold over seven tonnes of wrought iron. The structure's base is 125 metres across with a height of three hundred metres. The structure was conceived by Gustave Eiffel (1832-1923) for the 1889 Universal Exposition in Paris. It is a latticework grid pylon that was initially hated by the locals but is now seen as a national landmark. The reader can read for themselves more about this unique structure. The Louvre in Paris boasted an early example of a wrought-iron roof. The most photographed fence in the world has to be the fence surrounding Buckingham Palace in London. This Palace is the residence of the Queen of England and the Commonwealth. This wrought-iron fence was the work of the Bromsgrove Guild of Applied Arts, founded in 1898 by Walter Gilbert. The fence was erected in 1911 and remained a particular ornamental feature of the Palace. The following structure that I want to mention is the Bollman Truss Semi-Suspension Bridge in Maryland, the United States (mentioned previously under Chapter Two-Bridges). This bridge is a National Historic Landmark and the last surviving wrought-iron bridge in the world. The last wrought-iron structure I want to discuss is The Iron Pillar of Delhi. This wrought-iron structure is over seven metres high. The Pillar is famous for its lack of rust and resistance to corrosion (the reader can find out more for themselves). The last structure that uses wrought iron is the Statue of Liberty, located in New York, the United States. This neoclassical structure was designed by a French sculptor called Frederic Auguste Bartholdi (1834-1904). The cast iron metal framework was by Gustave Eiffel (mentioned earlier). The statue is ninety-three metres high, with a torch in her left hand measuring nearly nine metres long. The statue was dedicated as a UNESCO World Heritage Site in 1984.

The second type of iron is cast iron. There are three different varieties: white cast-iron, grey cast-iron, and ductile cast-iron. Making cast iron is melting pig iron in a furnace with added quantities of limestone and carbon. The advantages of cast iron are its resistance to wear and tear, low melting point and resistance to rust (if coated). This material is excellent for cooking utensils and engine parts for cars. Cast-iron contains 3–5% carbon. It is used for pipes, valves and pumps. It is not as tough as steel, but it is cheaper than steel. Magnets can be made of iron and its alloys and compounds. The disadvantage of cast iron is the brittle nature of the material. When a material is brittle, the material can break with little plastic deformation (the bending of material into a new shape under force). However, I would like to name a few famous cast-iron structures for your information. In engineering, two structures vie for first place. The first structure is Ditherington Flax Mill, located just outside Shrewsbury. This building was built in 1797 and is the oldest iron-framed building globally. Although the building is only five storeys high, it has been recognised as the world's first skyscraper. The building is now

Listed as a Grade 1 structure. The Iron Bridge at Coalbrookdale, Shropshire, is the second structure that vies for first place. This bridge was built in 1777 and was completed in 1779. A local architect from Shrewsbury first designed the bridge called Thomas Farnolls Pritchard (1723-77). There were 378 tons of iron used to cross the thirty-metre gap of the Ironbridge gorge. This bridge was the first structure ever to be made of cast iron. The iron for the bridge was cast by Abraham Darby III, a relative of the name mentioned earlier. Another structure that uses cast iron is the Pontcysyllte Aqueduct. This aqueduct is a navigable aqueduct that carries the Llangollen Canal over the valley of the River Dee in Wrexham, North Wales. The aqueduct has eighteen hollow piers that are slimmer at the top than at the bottom, which rises to 125 ft high. There are approximately 19 arches that have a 45ft span between them. The structure is over a thousand feet long and is the largest and longest aqueduct in Britain. A cast-iron trough carries water over the aqueduct, which is over eleven feet wide and five feet deep. It was started in 1795 and completed in 1805, and the engineer was Thomas Telford (born 1957, died 1834). He was a famous Scottish civil engineer and became the first president of the Institution of Civil Engineers. Other structures use cast iron; however, the reader can find out more for themselves.

I have included iron in this book because this was an original product, and our world would look very different without this material. As manufacturing processes evolved, iron was cast aside in favour of an iron alloy, steel, and that will be mentioned in a later chapter.

CHAPTER TEN – THE LETTER J

The option chosen for the letter J is the SI Unit for energy called the Joule. A joule is defined as "the energy transferred to (or work done) an object when a force of one newton acts on that object in the direction of the force's motion through a distance of one metre". The Joule is named after James Prescott Joule (1818-89), an English physicist, mathematician and brewer. His work on the nature of heat and energy led to the development of the first Law of thermodynamics (the total energy of an isolated system is constant; energy can be transformed from one form to another but can neither be created or destroyed). Let us consider an example. In modern buildings, insulation is required to retain the heat inside the building. In other words, the building will retain some of the heat, lose some of the heat through windows (it should be noted that all houses lose heat), and the remaining heat will have to top up by using central heating. Therefore over the years, as a Western population, we have added double glazed windows to our houses (we will discuss heat loss through building materials, the U Value, measured in Watts per square metre, per degree Kelvin, W/m2K, in another chapter). Therefore, as heat is a form of energy, the energy is not "lost"; it is relocated to the outside of the house. I hope this helps!

Before starting this chapter in earnest, I would like to discuss the previous statements. A newton is the SI unit of force, named after Sir Isaac Newton (mentioned earlier). The newton is the force needed to accelerate one kilogram of mass at one metre per second squared. The units per metre per second squared refers to velocity, the idea that an object is moving at speed (gravity is the force pushing down on a person, and gravity is known as 9.81 metres per second squared). A joule, as defined above, is a unit of energy. According to the Internet, one Joule is the energy required to lift a 100-gram object one metre in the air. There is another example that I would like to include. One Joule is equivalent to producing one watt of power for one second. A kilo-watt hour is equivalent to 3.6 megajoules (3,600,000 joules). We will discuss these in a later part of the chapter.

Kinetic energy is defined in physics as the energy that an object possesses when in motion, and the standard unit is the Joule. A good example of kinetic energy is a river flowing, as mass and volume flow at a certain speed. Another example would be the large amount of kinetic energy of an asteroid falling through space. Another example of kinetic energy is ice cream. The invention of the ice cream freezing machine by Nancy Johnson (1795-1890) gave ice cream to the whole world. Due to the warm air surrounding an ice cream, the particles within the ice cream move faster, causing the ice cream to melt. Therefore, melting ice cream is an example of thermal energy.

Potential energy is another form of energy, stored energy from the object's position relative to its surroundings. An example would be the wrecking ball on a crane. When the ball is in its usual position, there is no energy. However, when the ball is elevated to start its swing, there is potential energy as the ball gains height. There are four forms of potential energy that I would like to discuss. Gravitational potential energy, or mechanical energy, can be defined as an object's potential energy concerning another object due to its height, known as gravity. There are a couple of examples that I would like to mention. We mentioned in the previous chapter dams. The water behind that dam has potential energy, as the water behind the dam is higher than the water in front of the dam. When the water level lowers, and the water is used to turn turbines to create electricity, the energy is converted into kinetic energy. I want the reader to discover more for themselves, as hydroelectricity is a fascinating subject. There is another form of potential energy, and that is chemical energy. This energy can be described as the energy that exists in an element due to the bonds between the molecules and atoms that make up the element. In other words, a chemical element transforms into another object when, for example, heat is applied. The atoms and molecules are re-arranged in the new element. This transformation is the only point when chemical energy can be measured. Let us look at an example. We have mentioned in previous chapters fossil fuels and how they are used. Wood and coal are both burnt, creating heat and light. Another example that every child should know is photosynthesis, when light shining on a plant (solar energy) is converted into chemical energy (in this case, glucose inside the plant). There is another form of potential energy, and this is electrical energy. The concept of potential electric energy is defined as the stored energy that a particle has before it is set in motion by an electrostatic charge. An example is a headlight bulb on a car fixed to a battery. The car battery has both a negative and positive charge. Electric potential energy is the energy stored in the electric circuit before the headlights are switched on. Another form of potential energy is radiant energy. This energy is the energy of electromagnetic and gravitational radiation. Electromagnetic radiation refers to radio waves, microwaves, X-rays and visible and non-visible light. Gravitational radiation refers to gravity waves that occur in space, which we will not cover in this chapter. Radiant energy travels in waves and is everywhere, even though it is impossible to see these energy waves. A good example is radio waves transmitted through the air. The work was started by Heinrich Hertz (1857-1894), a German physicist, a man who proved the existence of electromagnetic waves. The Hertz (cycles per second), the SI unit of frequency, was named in his honour. This theory was first proposed by a man called James Clark Maxwell (1831-1879), a Scottish scientist of mathematical physics. His theory was to bring together the ideas of light, electricity and magnetism into one field. The work was published as "A Dynamical Theory of Electromagnetic Field" in 1865. The theory states that electricity and magnetic fields travel through space as waves move at light's speed. This theory led to the production of radio waves and started the movement of modern physics. The work that was undertaken by Marie Curie (1867-1934), a Nobel Prize Winner in both physics and chemistry, is worth investigating. There are other forms of energy, such as elastic energy and nuclear energy, and the reader can find out more for themselves.

In the last part of this chapter, I would like to discuss some different magnitudes of joules and give some examples that I hope you will find interesting.

In a previous chapter on gunpowder and the effect, this had on the world. In measuring the destructive force of an explosion, the force is measured in terms of the energy released. The measure used is a ton of TNT (trinitrotoluene, a compound with C6H2NO2CH3). A ton of TNT is a unit of energy equal to 4.184 gigajoules (4.184 x109J). A kiloton of TNT is a unit of energy equal to 4.184 terajoules (4.184 x1012J). A megaton of TNT is equivalent to 4.184 petajoules (4.184 x1015J). I want to leave this part here as I will be discussing the subject in another volume.

Let us now consider some more admirable examples of energy used! A yoctojoule (10-24) is one million times four smaller than a joule of energy. It is recorded on the Internet that 1.6 yoctojoule is the energy of a typical microwave oven photon (the reader can look this up for themselves). A femtojoule is (10-15) of a joule and is the kinetic energy of one human red blood cell. A nano-joule is one billionth (10-9) of one Joule. It is recorded on the Internet that 160 nano-joules are the kinetic energy of a flying mosquito. The use of laser technology to replace the eyes' natural lenses is a twentieth-century phenomenon. The laser operates at a level of one nano-joule, and I will be discussing lasers in a later volume. A Decajoule (102) is a hundred times greater than a joule, and three times this amount of joules is a lethal dose of X-Ray rays. A Kilojoule is one thousand times greater than a joule (103). A square metre of Earth receives 1.4 kilojoules of solar radiation during daylight hours. More energy from the Sun hits Earth every hour than the entire planet uses every year! A megajoule is equivalent to one million joules (106). One kilowatt-hour is equivalent to 3.6 megajoules. According to the Internet, ten megajoules of energy are used every time a person takes a ten-minute shower. It takes a hundred megajoules to supply an average house with electricity for one day. A petajoule (1015) is a thousand million million joules. In Australia, eight hundred petajoules are Australia's natural gas consumption. The energy released by the volcano Krakatoa in 1883 was eight times the power of a joule to one hundred quadrillions (this is equivalent to 8 x 1017J). To heat all the water in the world by one-degree centigrade, it would take one yottajoule (1024), a septillion joules. The last one will genuinely amaze me; the total mass energy of our Milky Way (our galaxy) is one hundred octadecillion (1x1059J).

I hope this chapter has proved interesting for the reader. There is plenty of information for further research, should you wish to do so.

CHAPTER ELEVEN – THE LETTER K

The option for the letter K is Kelvin, a temperature scale that is mainly unheard of by the public. Temperature is the expression used to discuss thermal energy mentioned earlier. Many people will be used to the Celsius and the Fahrenheit scale of temperature (there is a third temperature scale, known as the Rankine scale, but we will not pursue this further here). By way of explanation, both the Celsius scale and the Fahrenheit scale work very well on the Earth's surface. Both scales are measured at standard atmospheric pressure at sea level, which is why they are standard measures for everyday use. The problem occurs when we consider space, and I will elaborate further later in the chapter.

We will now discuss both the Celsius and Fahrenheit temperature scales before considering the Kelvin scale in depth. The author knew the Celsius scale as the centigrade scale was named after the Swedish astronomer Anders Celsius (1701-44). By way of the explanation, the centigrade scale comes from two Latin words, centum meaning 100 and gradus meaning steps. In 1742, Anders Celsius wrote a paper called "Observations of two persistent degrees on a thermometer", where he proposed that the boiling point of water is 0 degrees Celsius and that ice is formed at 100 degrees Celsius. Jean-Pierre Christin (1683-1755), a French physicist and astronomer, proposed in 1743 to invert the Celsius scale to what we understand today. By coincidence, Carl Linnaeus (1707-78, the famous botanist) in 1744 reversed the original Celsius scale. The term centigrade was abandoned in 1948 when Celsius was more formally used. As a pretty exciting piece of information, the Met Office in 1985 stopped using the term centigrade and started using the term Celsius. The Fahrenheit scale was named after Daniel Gabriel Fahrenheit, a German physicist (1686-1736). Daniel Fahrenheit started his theory on temperature by considering the average human body temperature (initially 960F; however, this has been revised since then to 98.60F). There are two points to be remembered when considering the Fahrenheit scale. At 320F pure water converts into ice, and at 2120F is the boiling point of water. In 1724, Daniel Fahrenheit, a glassblower by trade, proposed a system for making thermometers and the Fahrenheit temperature scale in a paper to the Royal Society of London. Daniel Fahrenheit was the first person to put mercury inside a thermometer. Daniel Fahrenheit was elected a Fellow of the Royal Society in 1724.

A Scot named William Thomson, 1st Baron Kelvin (1824-1907), discovered the Kelvin scale. The research work undertaken in physics and thermodynamics significantly accelerated our knowledge of how temperature works. The knighthood in 1892 was bestowed upon him for his work in thermodynamics. The Kelvin scale does not have negative numbers, which means it is famous for measuring liquids, which we will discuss later. There are no negative numbers, which means that 0K is equivalent to -2730C; this is the lowest form of temperature at which life can exist (there are no degrees of temperature in the Kelvin scale, just numbers).

In order to understand the importance of the Kelvin scale, there is a practical example to discuss. We would all like to use the central heating system less in our proposed modern sustainable world. Not only does this cost less money, but it is better for the planet. Many people who live in modern houses have double glazed windows. There is a good reason, and that is to improve the thermal insulation of a house. Therefore, when discussing building temperatures, we always use the Kelvin temperature scale. Three measures require explanation. The Lambda Value (λ) means the rate at which heat is transferred through a material (thermal conductivity). Fourier's Law is named after a French physicist, Joseph Fourier (1768-1830). Fourier discovered that certain materials allow heat to pass through them more straightforward than others. So copper allows heat to pass through easier than steel, for example. The Lambda value is measured in Watts per square metre of area for a temperature of one

Kelvin per metre thickness (W/mK). We use the Lambda value when considering U-Values, which are mentioned later in the book.

I would now like to have a lighter moment when we consider such a heavy subject. The author would like to discuss two planets that orbit the Sun, Mercury and Venus. A fun fact for the reader is that ninety-nine per cent of our solar system mass is the Sun itself.

The first planet is Mercury. This planet is a minor planet in our solar system, and the orbit around the Sun takes nearly eighty-eight days. The planet was formed over four billion years ago. There is no atmosphere at ground level on Mercury, as the Sun's rays are too strong. The temperature range of Mercury ranges from 8000F (4270C or 700K) to minus 2750F (minus 1700C or 103K). We mentioned earlier in the book the "wobble" in the axis of Mercury around the Sun. Albert Einstein explained this theory; however, there is a more straightforward practical example. Some believe that an asteroid hit the planet, and there is evidence of a large crater on the surface, which may have affected the planet. It is now believed that the planet rotates on its axis three times for every two orbits around the Sun (I would suggest the reader consult a computer for a more detailed explanation).

The next planet is Venus, the second planet from the Sun. The planet orbits the Sun every 224 days, and at certain times the planet can be visible from Earth. The size of Venus is comparable to Earth, which is why it has been called the "sister" planet. However, there are some dramatic differences. There is a ninety-six per cent carbon dioxide atmosphere on Venus (on Earth, there is 0.04% carbon dioxide with nitrogen at 78% being the most numerous gas). The atmosphere on Venus is ninety-two times greater than Earth. This difference means that a person will feel the same pressure as if they were miles under the sea. One of the most amazing facts about Venus is that it rotates opposite to Earth. The last fact that I want to share is the temperature. The average surface temperature on the planet's surface is 4710C (equals 744 Kelvins). This temperature makes Venus the hottest planet in our solar system.

The last fact that I will leave you with is that Mercury and Venus are the only two planets that do not have moons. For example, I could discuss all the other planets in the solar system, including Uranus, Jupiter, and Mars. For instance, the Olympus Mons on Mars is the highest mountain ever discovered, and I want the reader to discover more for themselves. Space is a fascinating place, with facts that will blow your mind.

I hope this chapter on the Kelvin temperature scale has filled you with enthusiasm for more research!

CHAPTER TWELVE – THE LETTER L

The option is chosen for the letter L is light, part of the electromagnetic spectrum. Before considering the various practical uses of the electromagnetic spectrum, we will discuss the basics of light. We use electricity and light bulbs to light our way in the modern world. I want to finish the chapter with a natural wonder known as the Northern Lights.

So, let us first discuss light in general. The electromagnetic spectrum is an all-encompassing term for all the frequencies of electromagnetic radiation. The Gamma wave has the shortest wavelength, ten picometres (one trillionth of a metre). The unit of a picometre is used for measuring atoms. The next shortest wavelength is X-rays, which have a wavelength up to 1 nanometre (one billionth of a metre). Gamma waves and X-Rays can enter the human body and cause potential damage to the fabric of cells within the human body. This process is called ionisation. The next part of the spectrum is microwaves which are not visible, with a wavelength of one millimetre to one metre in length. Radio waves have a wavelength of one millimetre to a hundred kilometres. Wavelengths visible to the human eye range from ultraviolet (100 nanometres) to infra-red (One micrometre, one-millionth of a metre). Light, both visible and invisible and radio waves cannot enter the body and are called non-ionising rays.

I want to discuss now, albeit briefly, two different wavelengths of electromagnetic radiation. Gamma radiation was discovered by Paul Villard (1860-1934), a French scientist and physicist examining the radiation waves from radium. Gamma rays have the shortest wavelength; however, they have the highest energy of any form of light. An artificial means of creating gamma rays are in a nuclear reactor, and research into Gamma rays took part as part of the Manhatten Project (the nuclear bomb). Gamma rays occur in space with the death of stars (but that is beyond the scope of this book). However, there are practical uses in medicine, using gamma rays to kill specific cancer cells. The most famous (fictional) example of a medical experiment with gamma rays has to be Bruce Banner turning into the Incredible Hulk, as portrayed in the Marvel Comics (however, that level of radiation poisoning in real life would have killed him). In older literature, the difference between X-rays and gamma rays was seen as the difference in wavelength between the two light waves. However, the modern difference is that gamma waves occur from a reaction inside a material, and X-rays occur outside the nucleus of a material. X-rays were discovered by Professor Wilhelm Roentgen (1845-1923) in 1895 by accident as he was busy considering how cathode rays could travel through glass. His accidental discovery awarded him the first Nobel Prize in physics in 1901. X-rays are primarily used in medicine as they travel through the human body and highlight tissue and bone damage. The images are produced when part of a person's body is placed between the X-ray machine and an X-ray photographic plate that detects the image. As bones absorb more radiation than the surrounding tissue, an X-ray image is produced on a film. However, Elizabeth Fleischman (1867-1905) should be mentioned as an X-ray pioneer. The reader should remember that there was no protection used when X-rays were being produced in the beginning. Consequently, this lady and other unfortunate people suffered severe side effects such as burning and ulcers on their hands after repeatedly demonstrating an X-ray machine. However, Elizabeth Fleischman continued her work as a radiographer and recommended protection measures like lead and double-plated glass. Unfortunately, her pioneering work led to her death from lung cancer at 38. In the book Pythagoras in the Corner, I mention Clarence Dally, the first man to die of radiation poisoning. These pioneers' recommended measures were finally introduced, and now X-rays are taken safely throughout the world. Microwaves and radio waves are part of the electromagnetic spectrum and are used in radar systems (more on later). There are, of course, many other forms of electromagnetic radiation, and I would recommend the reader find out more for themselves.

Any discussion regarding light should include electricity. As part of our modern world, electricity provides most of the light we require to live. The word electricity comes from the Greek word elektron, meaning amber. Many inventors tried to create the first light bulb. A Canadian patent was filed in 1874 by two inventors called Henry Woodward and Matthew Evans for the light bulb that used carbon rods; they were unsuccessful in making this a commercial enterprise and sold their invention to Thomas Edison in 1879. Thomas Edison (born 1847, died 1931) was an American inventor and businessman with over one thousand patents. He is credited with inventing the motion picture camera and the incandescent light bulb and designed the sockets, switches, fuses and meters that we use today. Thomas Edison improved on the work of previous inventors by using tungsten as a filament in lamps, thereby creating a bulb that would last longer than earlier versions. However popular these old-fashioned lights, they emit more heat than light. The future of the lightbulb lies with compact fluorescent light bulbs, called CFLs. Edward E Hammer invented these in 1976 in response to the oil crisis in 1973. However, due to the high manufacturing cost, the product was shelved. Other companies copied this design, and since 1980, it has become widely available. They are made from glass moulded into a shape, the most common being either spiral or standard, which looks the same as an ordinary incandescent light bulb. The bulb's interior is coated with white phosphorous, and the oxygen inside the tube is pumped out and replaced with mercury gas. The tube is sealed and a base is added to use the appliance in an ordinary household socket. So what are the advantages and disadvantages of energy-efficient light bulbs? The advantage of using energy-efficient light bulbs is that they use about one-fifth of the energy of a standard incandescent light bulb. This action means they are far more efficient in lighting the room, as less electricity is required to light an energy-efficient bulb. A good way of knowing which CFL bulb to choose is to divide the wattage required for an incandescent light bulb by four/or five and use the CFL bulb you require. So a standard 60 watt incandescent light bulb would only require an 11-14W CFL. This smaller wattage naturally results in lower household electricity bills by about ten per cent per year. The other significant advantage to a CFL is that it will last approximately ten to twenty times longer than a typical incandescent light bulb, depending on usage. The higher cost of a CFL bulb is more than offset by how long the bulbs last. There are some disadvantages with CFL bulbs, as they contain mercury. This substance can damage householders and require a careful clean up if the bulb is broken. You will have to leave the room and wait for about ten minutes till the air is clear. Mercury is particularly dangerous as it attacks the central nervous system; symptoms include tremors, emotional liability, memory changes and neuromuscular changes. It also attacks the kidneys and lungs of those that handle mercury. It can lead to stillborn births in pregnant women. CFL bulbs cannot be disposed of in the ordinary method like ordinary bulbs due to their mercury. It is recommended that they do not go into landfill and are disposed of at a recognised centre. However, mercury has been significantly reduced in modern CFL bulbs and maybe down to as little as 0.4 per cent in modern lighting.

I want to discuss with the reader the natural phenomenon known as the Aurora Borealis, or Northern Lights. The lights in the southern hemisphere are known as the Aurora Australis or Southern Lights. Since ancient times, these Lights have been observed, and many legends have been told. The Vikings thought the Northern Lights were caused by the glimmer of weapons held by immortal warriors. The Alaska Inuit people thought that these Lights were the souls of salmon, deer and other animals. The Northern Lights have been recorded by Galileo, Benjamin Franklin and Edmund Halley (Halley's comet). The most incredible show of lights happened in 1859 when there was a tremendous electromagnetic storm. The lights are caused by electrically charged electrons and protons coming from the surface of the Sun and hitting the Earth's magnetic fields. These protons and electrons are carried by solar winds (a solar wind is a stream of electrically charged particles emitted from the Sun). Since these particles are charged, they travel to the Earth in a spiral formation along magnetic lines, converging at the North and South Pole. The Northern (and Southern) Lights are formed when these charged particles from the Sun mix with the gases found naturally within the Earth's atmosphere. These gases then glow, causing the Lights to appear. The Earth's atmosphere is made of nitrogen and oxygen, with oxygen atoms at distances up to six hundred miles above the Earth. The Lights dance above the Earth in colours ranging from green to red and orange. Oxygen atoms emit green and red light, whereas nitrogen atoms emit more orange and red light. The Auroras are visible from space, and many astronauts onboard the International Space Station have taken pictures from space. Satellites have also taken pictures of the Auroras, visually stunning photographs. In January 2015, NASA launched a rocket known as the Auroral Spatial Structures Probe (ASSP) to study electromagnetic energy in the atmosphere. It should be noted that activity on the Sun is closely monitored by space agencies on Earth, such as NASA, as electromagnetic storms can disrupt both telecommunications equipment and geospatial satellites.

I hope this chapter on light has been illuminating (sorry, I could not resist!).

CHAPTER THIRTEEN – THE LETTER M

The option chosen for the letter M is maritime. This subject is vast, and I realise that I will not be able to cover all aspects. The first part will look at the different sections of a boat and what they are called (this is easily found on the Internet; however, I will give a brief resume here). Next, I will look at the differences between a boat and a ship. At this point, I will try and give a brief history before we look at the inventions that made modern boats and ships. Lastly, there is one ship that I will mention that was one of the worst maritime disasters of all time.

The front part of a boat is called the bow, and the rear is called the stern. The transom is the stern cross-section of the boat. The hull of a boat is a part of the boat that is underwater. Logically, it will be watertight and is generally used for carrying goods and passengers. The line where the hull meets the water is called the waterline. The draft of a boat is the distance that the boat requires to float on water, measured from the waterline to the bottom of the boat. The term gunwale refers to the top edges of a boat. The beam of a boat refers to its widest point. The last two terms that I will refer to here are port and starboard. If you cannot remember which is which, remember that both port and left are four letters, this should help! (As an aside, the acronym POSH, generally meaning wealthy, has been commonly referred to as Port Out Starboard Home. As ships were heading for India, the wealthiest preferred to be on the shadiest part of the boat on the outward journey, which was on the left-hand side. Therefore, the homeward bound journey is on the starboard, or right side of the boat. There is no evidence for this; however, it has become part of our daily language).

There are several differences between a boat and a ship. The first difference is size; a ship is generally more prominent than a boat (a ship can carry a boat; however, a boat cannot carry a ship). This phrase is vague but accurate. Another point to note is the "three-plus square-rigged masts" rule. No boat can carry more than three masts; otherwise, it would be too slow. On the other hand, a ship can have more than three masts. A boat can be powered manually using oars; a ship requires an engine due to its size. Another difference between a boat and a ship is the distance travelled. A boat will generally stick to local areas, whereas a ship can travel longer distances over oceans. The final difference that I would like to mention is the use of a vessel. Boats can be used for recreational purposes, whereas a ship is only ever used commercially. I hope that this has helped the reader.

The history of sailing stretches as far back as history can remember. In the beginning, rafts were used of wooden members tied together to form a floating platform. There were even tree trunks that were hollowed out and floated. In New Zealand, there are examples of Maori war canoes. The most famous is Nga Toki Matawhaorua (the reader can find out more for themselves), which the author has seen, the largest canoe in New Zealand. The canoe is over thirty-five metres (123 feet) long and two metres wide (six feet). However, it should be remembered that canoes could only travel as far as the crew were able using oars. The most crucial invention in maritime history was the use of the sail,

and the central area of trade was the Mediterranean Sea. The sail enabled boats to travel longer distances and carry heavier loads. The pioneers of sea travel were the Phoenicians from modern-day Lebanon. They traded in the Mediterranean area between 1000 and 2000 BC. The trading of goods enabled the flow of knowledge from Egypt to Greece and other countries. One of the largest ancient Empires the world has ever known was the Roman Empire. The Roman Empire was founded in 27 BC by Emperor Augustus and was in existence till its fall in 467 AD. The Roman Empire stretched across three continents and was in existence for over five hundred years. This Empire completely encircled the Mediterranean Sea, and therefore the transport of goods by water was quicker than by land. The next seafaring race that I would like to discuss, and the reader may find interesting, is the Norse people, commonly known as the Vikings. This race of Norsemen came from Norway, Denmark and Sweden. Their mode of transport was known as longships, with documented archaeological evidence dating back to the fourth century BC. The longships were manually powered with oars, and sails were added later in the eighth century. The most common aspect of all the ships was that they were clinker-built (where the planks overlap each other and are butted against each other is called carvel construction). All of the longships had a shallow-draft hull to be used in shallow waters. Another common feature was that all ships were double-ended so that the sailors could see all around the longship. Due to their skill as shipbuilders, Western Europe was invaded, from France to Ireland and England. The Vikings' impact on Europe cannot be underestimated, as their language, trade, technology, and society left a lasting impact (the reader can find out more for themselves). The Age of Discovery, from the fifteenth century onwards (the reader can find out more for themselves), is enough to fill several volumes of books. Someone should be mentioned, called Sir Josepth Isherwood (1870-1937). He invented the Isherwood System of longitudinal construction for ships, which strengthened the keel of a ship. However, at this point, I would like to leave sailing history and discuss the great inventions that enabled new worlds to be discovered.

The first great invention was the marine chronometer. The direction of a ship is fundamental, and navigation was always an issue since sailing was first attempted. The early travellers navigated by way of astronomy, by looking at the stars above. It was seen as reasonably straightforward to know the latitude of where a ship was located (latitude is the north-south of the equator). A ship could be sure of this location using a sextant and a compass. However, the longitude position was always an issue. The boat's position generally relied on either two known land-based points, a compass bearing to a known point or a celestial bearing. There is a long history of chronometers; however, John Harrison (1693-1776), an English clockmaker, solved the problem. His timepieces with weighted springs helped ship navigators assess their location, using time-related to Greenwich Mean Time (the reader can find out more for themselves).

The second great invention that assisted sea travel was the internal combustion engine (discussed earlier). Benjamin Bradley (born 1830, death unknown) was born into slavery in the United States, and he was the first to develop a working model for a steam engine that could be used for ships. The first steamships and later diesel engines have enabled ships to travel around oceans.

The third great invention I would like to discuss is the screw propeller. Benjamin Montgomery (1819-1877), an enslaved person born in Virginia, invented the steamboat propeller. However, he was not allowed to patent his invention due to his status as an enslaved person, which is unjust in the author's opinion. However, it was the screw propeller that enabled ships to travel faster than before. There were two inventors of the screw propeller. The first, John Ericsson (1803-89), was a Swedish-American inventor who held US Patent No. 588. Ericsson designed the US Navy's first screw propelled ship, the USS Princeton. The other inventor was Sir Francis Pettit Smith (1808-74), who assisted in designing the first screw propelled steamship, the SS Archimedes. The reader is recommended to read about propellers, increasing their engineering knowledge. The engineering principle is straightforward, and the propeller rotates, causing the ship to move forward. The thrust and torque (mentioned earlier) of a propeller are created by the pressure differences on either side of the blades as they rotate.

Another invention that has been important both on land and at sea is the radar. No single person or country can lay claim to inventing radar. However, I will only concentrate on the United Kingdom concerning radar waves. The British people should be indebted to one person Robert Watson-Watt (1892-1973), a pioneer in radar technology. In the 1920s, Robert Watson-Watt concentrated on finding the location of thunderstorms by measuring the interference given by lightning to radio waves. Mr Watson-Watt suggested to the government that radar might be used to counter any threat from the bombing that might occur. In 1935, an experiment conducted by Robert Watson-Watt demonstrated that radio signals from a radio transmitter bounced off an aircraft travelling in the sky above. An experimental seaplane station was based at Felixstowe, and the aircraft was tracked for the first time. The 17 June 1935 was the birthday of the radar in this country. This wavelength was called pulse radar, whereby the time taken for the radio signal to return to the radar station could be measured. Radio waves travel at a known speed, the speed of light, which is 186,000 miles (300,000 kilometres) per second. For example, if the signal from the aircraft bounces back to the station 1/1000 of a second, then the plane must have a round trip of 186 miles, which makes the plane 93 miles away. The same technology is used in marine radar systems to navigate a vessel and avoid collisions. Microwaves are sent out from the antenna of a ship, and any obstructions such as ships bounce back to the antenna. These obstructions are then displayed on a screen inside the ship.

The latest invention to discuss is GPS (Global Positioning System). There is much information on the Internet, and I mean a lot. However, I will try and give a brief resume here. The GPS consists of over thirty satellites that encircle the Earth, which send signals to ground receivers (radar stations). These stations then relay that information to your phone or other devices. However, there is not one signal coming from these satellites; there are three or four signals. By a method known as triangulation, your exact location on Earth is given in the form of coordinates in a longitude and latitude position. There is an incredible history behind satellites and the GPS. Several countries have satellites above Earth that are used for GPS. These are known as GNSS (Global Navigation Satellite Systems), and I will let the reader find out more for themselves.

Later on in the book, there is a discussion on the famous Liberty ships that helped the Allies win the Second World War.

A famous ship should be included in this chapter. The ship in question is the RMS Titanic (Royal Mail Ship (Steamer), a boat allowed to carry post for the Royal Mail). This ship sank on its maiden voyage to New York on 14 April 1911 due to hitting an iceberg. Over 1500 people died in the disaster. This event was one of the greatest maritime disasters of all time. Many bad design decisions were made before the ship even left port, which would spell eventual disaster. There were insufficient lifeboats for the number of people aboard the ship. The watertight compartments meant to show that the ship was unsinkable did not work. Water could penetrate from one compartment to the next. The other fault was to do with the rivets that were used to put the ship together. It was later discovered that the rivets were substandard. This bad design meant that the rivets would pop open should the ship suffer a collision (which it did). An interesting story concerns the RMS Republic. In 1903, the RMS Republic was involved in a collision with the SS Florida's Italian steamer. It was rumoured that there was a large amount of bullion carried with the ship, although there have been no successful attempts to retrieve the bullion. As a result of the collision, six people died; however, over 1500 lives were saved. This action was due to the Marconi radio equipment that was placed on board and meant that ships in the area could respond to the distress calls that the ship gave. This event was the first time a distress call CQD (the SOS call was used later) was used by a naval vessel. The Marconi radio operator, a man called Jack Binns, was hailed a hero. Due to his heroic efforts, he was to be placed on the next ship that was to be built by the White Star Line, the RMS Titanic. The offer was withdrawn as it felt that he would bring bad luck to the voyage! (Jack Binns later went on to great success in business and died in 1959). Incidentally, the captain of the RMS Republic was not so fortunate as Captain E J Smith was Captain of the RMS Titanic.

I hope this chapter has been interesting for the reader. There was so much information to give, and I wanted to show the best information in this volume. God willing, I hope to produce another edition later.

CHAPTER FOURTEEN – THE LETTER N

The option chosen for the letter N is nickel. This metal may seem a strange choice, but nickel is a metal with an interesting story. In order to understand nickel, we will briefly discuss the history of this metal. The author will discuss the two major factors that make nickel such a valuable metal. At the end of this chapter, I will discuss the planet's heart, made of nickel, which will be examined below.

In 1751, Baron Axel Frederik Cronstedt (1722-65), a Swedish mineralogist and chemist (considered the father of mineralogy), found a new silvery metal from a mine at Los, in Sweden. The metal was initially considered copper, as copper and nickel are usually found in meteorites that fall to Earth (volcanic eruptions can also contain large amounts of nickel). Miners who found the material could not extract copper, so they nicknamed the material "kupfernickel", as it was thought (Old Nick, or the Devil), was present in the material. The material has been found in large deposits in the Sudbury mine in Canada, and the Norlisk mine in Siberia, Russia (one of the most polluted cities in the world). As an aside, the Sudbury mine was discovered during the building of the Canadian National Railway. However, the primary deposit of nickel is found in the core of the Earth, discussed later in the chapter.

The Bronze Age came after the Stone Age (as an aside, there is an animated film called Early Man, made in 2018, that deals with this!). This era was the next technological advancement of civilisation, and the next stage was the Iron Age. To put these into chronological order in metals is somewhat simplistic;

however, it helps our understanding of that period (the reader can find out more for themselves). There have always been reasons why countries invade each other. Before the advent of the oil and petroleum industries, metal was seen as a prime reason for the invasion. When the Romans invaded Britain in AD43, the metal tin was a significant asset. By combining tin with copper, we produce the alloy bronze.

However, we need to discuss what happens when the iron is alloyed with other elements, including nickel. When nickel is combined with steel, there is a massive increase in the strength of the combined metals. Nickel is a munitions element that was strategically important in World War Two. Nickel has two significant factors that make the metal useful. Nickel plays such an essential part in the maritime industry due to its resistance to corrosion in seawater and the lack of bio-fouling. These factors affect a boat's performance, as these can lead to a loss of speed and expensive coatings that will need to be applied. The other major factor that makes nickel useful is the ability to withstand high temperatures. Nickel has been used in the aerospace industry in the turbine blades of jet engines (discussed earlier). There are examples of a nickel-steel alloy used in armour plating for tanks, anti-aircraft guns and marine engines. Lightweight and rigid portable bridges (pontoons) were partly made from nickel, enabling troops to move easier across any terrain. An evolution of armour plating might prove interesting for the reader, from Harvey armour to Krupp armour and modern synthetic materials (the reader can find out more for themselves). Tanks, firearms and ships have all been made with a nickel alloy. As mentioned before, nickel has two significant factors that make valuable metal in the modern world. Nickel plays such an essential part in the maritime industry due to resistance to corrosion in seawater and the lack of bio-fouling. In the 1970s, a shrimp trawler named the Copper Mariner was receiving attention from the maritime industry. The boat's hull (see earlier in the book) can be made from copper and nickel, Cupronickel (Cu-Ni) alloy. Due to the Cu-Ni hull of the boat, it was shown that there was minimal corrosion and bio-fouling of the ship's hull. It was estimated that the hull's manufacturing cost could be recuperated in less than seven years due to the increased performance of the hull and the lack of ongoing maintenance costs. These factors mean that nickel has been used in various harsh environments that would not be suitable for other softer materials. These uses have been in the chemical industry, such as petroleum refineries (discussed earlier). Medical equipment and cookware are made from stainless steel, containing nickel (and other elements discussed later in the book). Stainless steel makes the equipment easy to clean and sterilise. The last use that the author can find is coinage. Since first introduced, the US (United States) five-cent piece has been called a nickel. The five-cent coin typically contains twenty-five per cent nickel and seventy-five per cent copper, although this will change in the future. There are other advantages to nickel that form part of our modern world. Domestic solar heating consists of a solar panel located on the roof of a house, and the heated liquid from the solar energy is transferred from the panel and installed into a storage tank for immediate use. Batteries may also be required; these are different from car batteries; they are called deep-cycle batteries and are either lead-acid batteries or nickel-cadmium batteries, both of which require careful storage due to their high risk. The water heated by using solar energy will help the life expectancy of your boiler, as the water is already heated by the time it reaches the boiler, thereby creating energy efficiencies. Solar panels work by converting sunlight into electric power on a DC (Direct Current) system. In domestic property, the Direct Current is transformed into Alternate Current using an inverter. This current is used to heat the house and the water used. Before exporting back into the Grid system, a meter will measure how much energy your house uses. Any excess power produced is fed back into the National Grid, where the credit will appear on your electricity bill.

In the last section, I will discuss the Earth's core. For simplicity, I have divided the Earth into four different main layers. There is an inner core, an outer core, the Mantle and the Crust. The inner core is solid iron and nickel. The inner core is located nearly six thousand kilometres (four thousand miles) beneath the Earth. The temperature reaches over 5,700K (about 5,4000C, 9,0000F) and is three million times the air pressure on you at sea level (330 to 360 gigapascals). The outer core is a ball of molten iron and nickel metal and is located over three thousand kilometres (nearly four thousand miles) below the surface of the Earth. The outer core is heated by the radioactive decay of uranium and thorium elements. There is a range of temperatures, from the outer regions, 4,300K (over 40300C, 72800F), to the inner regions adjacent to the Inner Core, 6,000K (5,7300C, 10,4300F). Because of such high temperatures, as shown, the Outer Core has a low viscosity (). The next layer is the Mantle, which accounts for over eighty per cent of the Earth's volume. The Mantle is further divided into two sections, the Lower Mantle and the Upper Mantle. The lower Mantle is over three thousand kilometres (nearly 1800 miles) thick. The temperature of the lower Mantle can reach over 4,300K (40000C, 7,2300F). The upper Mantle is solid, and both the Mantle and the Crust make up a layer called the lithosphere. The thickness of the upper Mantle is measured from the bottom of the Crust to the top of the inner Mantle. The depth has been measured as over four hundred kilometres (250 miles). In the upper Mantle, the temperature can range between K (5000C-9000C, 9320F-16520F). The Mantle is generally made up of iron, magnesium and silicon. The top part of the Earth, the Crust, is made up of large plates, called Tectonic plates, which move due to their differing density, and the Mantle below is constantly moving.

I hope this chapter has proved attractive, and there is a vast range of academic literature available if you are interested.

CHAPTER FIFTEEN – THE LETTER O

The letter O in this chapter stands for Oil and the transportation of Oil. I would urge you, dear reader, to think again if you believe that this is not a suitable subject. There are three times when a character in the Bond movies uses a pipeline. In the 1971 movie "Diamonds Are Forever", James Bond is rescued from a pipeline by disabling a welding PIG (Pipeline Inspection Gauge) in the movie. In the 1987 movie "The Living Daylights", the villain Koskov is sent through the Iron Curtain (this was the description for the area between East and West Germany after the Second World War, the reader should find out more for themselves) by using the Trans-Siberian pipeline (discussed below). In the movie "The World Is Not Enough", produced in 1999, the main plot involves a fictional incomplete oil pipeline that runs from Baku, in Azerbaijan (mentioned earlier) to Turkey. If it is good enough for the Bond movies, it is good enough for me! We have mentioned Oil before in a previous chapter, so I would like to continue the exploration of this subject.

Let us now have a brief recap of a previous chapter. Oil is formed from organisms that survived in the oceans, the plankton and other creatures that give life to this planet. As these creatures died and fell to the bottom of the sea, their remains were buried. Over hundreds of thousands of years, the pressure exerted on these creatures turned them into crude oil. The nickname "black gold" comes from the colour and the variety of uses for this resource. The history of oil exploration spans the entire world, everywhere an ancient sea was located! The top five Oil-producing countries globally are the United

States of America, Russia, Saudi Arabia, Canada and Iraq. Let us look at these in detail. In the United States of America, discovering oil in Pennsylvania in 1859 started an oil rush that remained unabated. However, it is the other areas, called Texas and New Mexico, that I want to discuss now. This region is also known as the Permian Basin. This area is a shale oil and gas region that covers an area of over two million acres of land. In Saudia Arabia, the Ghawar oil field measures over eight thousand kilometres squared. This oil field accounts for over a third of oil production for Saudi Arabia and pumps out nearly four million barrels of oil a day! There are, of course, other oil fields, and it may be of interest to the reader to find out more for themselves.

There are many ways to transport Oil around the world. However, I will only concentrate on two methods in this chapter for simplicity. The two methods are by the sea and overland, and let us consider each of these in turn.

Oil tankers are still some of the giant ships in the world. Continuing the Bond theme, the Liparus is a fictional supertanker owned by the villainous shipping magnate Karl Stromberg (played by Curt Jurgens 1915-82). An old supertanker that the Shell company previously owned was offered. The offer was declined because the residual vapour from the Oil is both toxic and explosive (the gases include benzene and hydrocarbon gas)! The other problem was that the ship would sit too high in the water and not look like it contained several nuclear submarines. In the end, a model was used (this will be covered in another book). Oil tankers are classified according to their size, which is the deadweight (DWT) of the tanker. The size can vary from a few thousand metric tons to over half a million metric tons to the supertanker weight. One of the biggest oil tankers in the world is called the Supertanker Europe, which was featured on the programme Impossible Engineering (Series 5, Episode 11). There are four ULCC (Ultra Large Crude Carriers), and they are named after four continents, namely Europe, Africa, Oceania and Asia. These ships are double-hulled (see below) to prevent oil disasters (please see below). The topmost part of the hull is painted white to reflect sunlight. As mentioned earlier, all oil tankers are powered by diesel engines, which was the invention of Rudolf Diesel (1858-1913). One of the most powerful diesel engines globally (Wartsila-Sulzer RTA96-C) is a turbocharged two-stroke diesel engine used in oil tankers. An impressive set of numbers is attached, like the crankshaft weighing three hundred tons (the reader can find out more for themselves). These incredible engines help propel these huge ships around the world. Regarding this, a mention should be made of the naval engineer William Froude (1810-79), whose formulas regarding water resistance to shipping propulsion (the hull speed equation) and shipping stability are still relevant today. Another engineer that should be mentioned is Sir Francis Pettit Smith (1808-74), who was one of the co-inventors of the screw propeller (alongside Swedish American inventor John Ericsson (1803-89)), mentioned earlier. There is another engineer that I would like to mention, called Joseph Becker (1897-1973), whose work on the Rudderpropeller (a combined propulsion and steering system) won the Elmer A. Sperry Award in 2004 (the reader can find out more for themselves). Elmer Ambrose Sperry (1860-1930) was an American inventor whose work on marine navigation using gyroscopes is still in use today. The gyroscope, which has a fascinating history, should be investigated by the reader. As a last note, all oil tankers now have an inert gas system (an inert gas is a gas that contains less than eight per cent oxygen) to prevent combustion. There is much history concerning the evolution of the vessels available on the internet. I will give a very brief history for those who do not have access to a computer. The first oil tanker (Vaderland, or Fatherland) was built by the Palmers Shipbuilding and Iron company, based in Jarrow, County Durham in the North of England. The site at Jarrow occupied about a hundred acres and, at its height at the beginning of the twentieth century, employed over ten thousand men. However, despite building ships for both the Royal Navy and private owners, the company went bankrupt in the 1920s due to the Great Depression. The famine and poverty suffered in the area led to the famous Jarrow march to London (the reader can find out more for themselves). The first successful oil tanker was the Zoroaster, floated in 1878 and designed by the engineer Ludvig Nobel (mentioned elsewhere). The ship had some unique features, with Bessemer steel (mentioned in Pythagoras in the Corner, another book by the author) for the main ship and iron (Chapter Nine) for the hold that would contain Oil. The other required feature was ventilation due to the explosive fumes and ballast tanks as the load is liquid. There are other features that I have detailed above. I would recommend the reader find out more for themselves.

There is a reason why there is a double hull for ships carrying large crude oil. The Exxon Valdez oil disaster occurred in Prince William Sound, Alaska, on 24 March 1989. The oil tanker spilt over ten million gallons (US gallons) of oil and is considered one of the worst oil spills worldwide. This event is due to the damage caused to the natural environment, which affected over a thousand miles of coastline (over two thousand kilometres), and affected all marine life. The Exxon Valdez was a single-hull ship, which was a significant factor in the accident (the reader can find out more for themselves).

The second method of transferring Oil is using pipelines over land. Pipeline transfer can transport water, crude oil (petroleum), biofuels and gas. The pipelines used in oil transportation are made of plastic or steel and are buried underground. Natural gas is also transported through carbon steel pipelines and is first pressurised lightly to form Natural Gas Liquids. The supply system of crude Oil starts with individual wells and then carries the liquid to a central collection point. The pipeline then carries the liquid either to a dock or onto a processing plant, where it is developed into the end product. For people without the internet, there are oil pipelines on every continent, apart from Antarctica. There are so many that there is not enough space to cover them all.

There are, of course, advantages and disadvantages with every transport system. I would now like to take this opportunity of discussing both sides of the argument.

Pipeline transportation is an economically advantageous mode of transportation. The liquid transported through the system can flow continuously and is not affected by the weather. The liquid will get high reliability from point A to point B. Another advantage is that the pipeline can take shortcuts, making transportation distance shorter than conventional roads. A further advantage of using a pipeline, compared to conventional rail and road transport, is that heavily populated city centres (railway hubs) are avoided. This action avoids the risk of accidents in cities, with spills and fires and the release of hazardous chemicals. In theory, there are only two parts to a pipeline: the pipe itself and pumping stations to keep the Oil flowing by increasing the pressure of the flow. This equipment makes the system simple to understand and monitor. Pipelines are incredibly safe and reliable and have low energy consumption compared to other modes of transport.

However, there are disadvantages. The fixed costs of a pipeline are huge. The controversial Keystone XL pipeline in America (the reader can find out more for themselves) is estimated at over $7 billion (over five million pounds). The use of fossil fuels is also controversial, and that is why the Keystone project has been abandoned by the US Government (as of this date, March 2021). There are environmental consequences to all major infrastructure projects, and oil pipelines are no exception.

I hope this chapter has been of interest to the reader and offered more directions for further study.

CHAPTER SIXTEEN – THE LETTER P

The following letter in our alphabet is the letter P, and I would like to discuss the concept of pressure. We are, of course, discussing force over a defined area. The Standard International (SI) unit of pressure is measured in pascals (Pa), named after a French mathematician, physicist and inventor, Blaise Pascal (1623-1662). Pascals are measured in Newtons per metre squared. When we discuss Newton's, it should be remembered that it is a force due to gravity. A Pascal is enough to provide a force that will move one kilogram at one metre per second squared. This force is generally measured at the Earth's surface. A good example is the concrete cube test. This test involves taking random samples of concrete and putting the concrete into cubes. These were of a size that is generally 150mm wide by 150mm long by 15mm high. The cubes were made of steel, as they must withstand the weight of the concrete that was put inside of them. The concrete has to be tamped down about three times and ensure no bubbles in the concrete. These cubes are left to dry and then put into tanks of water. The cubes are taken out of the tank and labelled to know where the concrete was taken from on the house. These cubes were delivered for testing by a specialist company in a controlled environment called a laboratory. These cubes were tested after about seven days, and the rest of the cubes after 28 days as the concrete hardens the longer you leave to set. The testing method was called the load testing of concrete or compression; in other words, a load was applied to the cube until it gave way. When the cube finally gave way or crumbled, the pressure was recorded and should be within a particular set of numbers or parameters. If this was, then the concrete was considered acceptable, and the rest of the concrete was considered safe. If the concrete cube failed the test, then this would mean that the foundations would not support the structure on top. This test was to find out the density of the concrete delivered and whether it could withstand the weight of the house.

In this chapter, we need to look at an older version of pressure that all engineers will need to understand. We will also consider the four different types of pressure, atmospheric pressure, absolute pressure, differential pressure and overpressure. This exercise will extend the reader's understanding with practical examples.

An older version of pressure is called pounds per square inch (psi). This measure was used to describe stress and how particles within a material exert force upon each other. This pressure is the opposite of strain, which results in a material deforming (bending) when a force is exerted upon the material. An excellent example of psi is blood pressure within the human body (2.32psig/1.55psig, also known as 120/80 mmHg (millimetre of mercury). Another set of examples is an American football at 12.5 to 13 psi, a bicycle tire is 65 psi, and a car tyre has 32 psi. In the previous chapter, we discussed pipelines, which operate at over 1000 psi. When liquid or a gas is being pumped through a pipe, it must be kept at a constant pressure. This liquid must be compressed every 360 miles by a compressor station to keep the system flowing. The gas is compressed by increasing the actual pressure of the gas or liquid. The pipe enters the compressor station cooling tower, and the gas or liquid is filtered free from impurities. The gas or liquid is then increased in pressure using either a turbine, electric motor or a reciprocating engine. The gas or liquid then leaves the compressor station to continue its journey.

A type of pressure that I want to discuss is atmospheric pressure, often noted as pa, which is the pressure of the Earth's atmosphere. When you are standing outside in the air, the atmospheric pressure is the average force of the air above and around you pushing in on your body. This pressure is also known as barometric pressure, after the barometer. The average value for the atmospheric pressure at sea level is defined as 1 atmosphere or 1 atm. Given that this is an average of a physical quantity, the magnitude may change over time based on more precise measurement methods or possibly due to actual changes in the environment that could have a global impact on the average pressure of the atmosphere. Therefore 1 Pa = 1 N/m2, 1 bar = 10,000 Pa, therefore 1 atm ≈ 1.013 × 105 Pa = 1.013 bar = 1013 millibar. Atmospheric pressure on the Earth varies with the altitude of the land. Therefore, there is less pressure on the top of mountains than sea level. The reason why we, as humans, are not crushed by atmospheric pressure is due to the blood pressure within our bodies being equal to the atmospheric pressure around us. Let us consider a rubber sucker being forced onto a wall. It is difficult to remove the sucker as air has been removed as we attach the sucker. There is also atmospheric pressure operating on the sucker. Therefore, we need to use considerable force to remove the sucker. Other examples of atmospheric pressure include sucking a drink through a straw, air pressure in a car tyre, and a toy balloon with air pressure inside. A vacuum cleaner has a fan that creates low pressure inside the device. Consequently, air and dirt particles are sucked into the device.

There is another member of the scientific community that I would like to introduce. Sir Robert Boyle (1627-91) was a philosopher, physicist and inventor. Boyle is primarily considered one of the founders of modern chemistry and a pioneer of modern scientific methodology. In 1662, Robert Boyle discovered that the volume and pressure of gasses are inversely proportionate when held at a constant temperature. When volume rises, pressure drops, and vice versa. There are many real-world applications of this theory that we now call Boyle's Law. For example, a spray can have paint expelled from its nozzle under pressure within the can itself. Spray painting with compressed air is a time-saving device compared to manual brushwork. A person of interest is Harold Ransburg, who developed a method called electrostatic painting so that any object spray painted will be coated evenly. Another example is a medical syringe, which will administer medicine under the vacuum pressure within the instrument. A can of soda, such as Coca-cola (other brands are available), is made under pressurised conditions. The last example is called the bends, and it is this example that I would like to discuss next. A slow ascension is critical for any properly trained scuba diver to know when they are ascending from deep waters. Our bodies are built for and accustomed to living in the average pressure of our lower atmosphere. As a diver goes deeper underwater, that pressure begins to increase. Water is heavy, after all. With the increasing pressure causing a decrease in volume, nitrogen gasses begin to be absorbed by the diver's blood. When the diver begins his ascent, and the pressure lessens, these gas molecules expand back to their average volume. With a slow ascent or through a depressurisation chamber, those gasses can work their way back out of the bloodstream slowly and normally. Nevertheless, if the diver ascends too quickly, the blood in their veins becomes a foamy mess. The same thing that happens to a foamy soda happens to a diver's bloodstream during the bends. On top of that, any built-up nitrogen between the diver's joints will also expand, causing the diver to bend over (hence its name) in severe pain. In the worst cases, this sudden depressurisation of the body can kill a person instantly.

I hope this chapter has proved to be of interest to you.

CHAPTER SEVENTEEN – THE LETTER Q

The letter Q is, perhaps, one of the most difficult letters of the alphabet. In strict engineering terms, the terms quantum mechanics and quantum physics should be mentioned at this point. The author looked into both of these and dismissed them as too complicated, to be honest! However, to make up for this omission, I have chosen other options, as shown below. These options are brief, and I hope the reader forgives the author. I have not included all the options, as there may be a chance of a second book.

The first option that I have chosen for the letter Q is the QR (Quick Response) Code. This code is essentially a barcode, the same type of code that appears on food and other articles. There can only be a maximum of twenty numbers in a conventional barcode corresponding to the article in question. The barcode was invented in the 1950s, and the reader can find out more for themselves if they wish to. The QR code was invented by Masahiro Hara in 1994 for the automotive industry in Japan. The inventor Masahiro Hara, by the company called Denso Wave. According to the company website, this company specialises in making automatic data capture equipment, industrial robots, and other ancillary equipment (according to the company website). The QR Code can store over seven thousand pieces of information and consists of black squares arranged in a grid pattern on a white background. The QR code is now used throughout the entire world by translating different languages to track and trace apps within the NHS (National Health Service). The author first came across the use of QR codes when the programme "Impossible Engineering" discussed the construction of the roof at Kansai International Airport, Japan. By using QR codes, the intricate roof could be constructed in a shorter time frame than by ordinary methods. I will let the reader find out more for themselves.

The second option is the quasar, short for quasi-stellar radio sources. The name quasar was used due to its discovery in the 1850s when the galaxy was scanned using radio waves. When an object in the galaxy was found, radio waves would bounce back to Earth. The distance to the object could then be measured. A quasar is a compact area in the centre of some galaxies. A quasar is also one of the brightest objects in the universe and can be observed across the entire electromagnetic spectrum. They are believed to be powered by a giant black hole (a place in space where matter and light cannot escape if they fall in). A quasar called 3C 273 was the first quasar to be identified. By measuring the wavelengths emitted from this object, it was estimated that the quasar was two billion light-years away from Earth (9.5 x1012 km, or 5.9 x 1012 miles). Therefore, this makes the quasar a distance of 9.5 million quadrillion kilometres or 5.9 million quadrillion miles from Earth (we will discuss quadrillion below), which is a huge number! The luminosity of this object is perceived to be four trillion times more luminous than our Sun. The light emitted from this object has taken over 2.5 billion years to reach Earth. There is a great deal of academic research on the Internet, and I would recommend the reader find out more for themselves.

The last option that I have chosen is a quantity measure known as a quadrillion. One quadrillion is equivalent to a thousand million million (1,000,000,000,000,000 short scale notation) and is expressed as 10 to the power of 15. In order to make this quantity more interesting, I have put together some interesting facts for the reader. It is about 586 quadrillion miles from one end of the Milky Way to the other. The Niagara Falls are located at Niagara Gorge, on the Niagara River. There are three Falls that make up Niagara Falls and these are Horseshoe Falls, American Falls and the Bridal Veil Falls. In 210 years, Niagara Falls will have used over a quadrillion gallons of water. The Great Lakes in Canada have a volume of about six quadrillion gallons. Many other facts, such as a quadrillion seconds is equivalent to 32 million years. I want the reader to have an exercise to see what a quadrillion looks like in trade and transport. The following number beginning with a Q, is quintillion (a million trillion), a number so large that a practical application is pretty useless outside of a scientific term. The reader can find out more for themselves.

There are two other options for the letter Q that I will include in a possible further volume. The first option is quicksilver, another term for Mercury's element. The other option, which will be far more fun to discuss, is the Q department in the Bond films, which stands for Quartermasters Office. In the author's opinion, one of the most famous Qs in the Bond films is the actor Desmond Wilkinson Llewelyn (1914-99), who played Q 17 times between 1963-1999. I expect the reader will agree that discussing the gadgets used by a fictional super-spy will be a fascinating subject indeed.

This letter was by far the most difficult in the whole alphabet. However, in subsequent volumes and later editions, I hope that I will be able to elaborate on this letter.

CHAPTER EIGHTEEN – THE LETTER R

There was only one option that I wanted to use for the letter R, and that was radar (Radio Detection And Ranging). The reason why is that this invention has been helpful for the last two centuries. In the twentieth century, radar was used to help defend this country. We are now using radar for a grander purpose in the twenty-first century. However, we really should start at the beginning.

The work was started by Heinrich Hertz (1857-1894), a German physicist who proved the existence of electromagnetic waves. The Hertz (cycles per second), the SI unit of frequency, was named in his honour. This theory was first proposed by a man called James Clark Maxwell (1831-1879), a Scottish scientist of mathematical physics. His theory was to bring together the ideas of light, electricity and magnetism into one field. The work was published as "A Dynamical Theory of Electromagnetic Field" in 1865. The theory states that electricity and magnetic fields travel through space as waves move at light's speed. This theory led to the production of radio waves and started the movement of modern physics. James Clark Maxwell had the most significant influence on physics in the 20th century and was named the third greatest physicist, behind Albert Einstein and Isaac Newton. Albert Einstein (1879-1955), a theoretical physicist (E=mc2), stated on the anniversary of Maxwell's birth, "his work was the most profound and the most fruitful that the physics has experienced since the time of Isaac Newton."

I am guilty of bias, but I will only concentrate on the United Kingdom concerning radar waves. The first part of this chapter belongs to the twentieth century.

There is one man whom the British people should be indebted, and that was Robert Watson-Watt (1892-1973), a pioneer in radar technology. In the 1920s, Robert Watson-Watt concentrated on finding the location of thunderstorms by measuring the interference given by lightning to radio waves. This work was not to be continued as, in 1930, the threat of Nazism and Adolf Hitler loomed large on the horizon. It was reported in the papers that the Germans had built a "death ray", capable of killing thousands in city centres and being able to stop bombers in the sky above. This information was proved false, as the technology could not exist at the time. The government contacted Mr Watson-Watt, and the theory was disproved; however, he suggested that radar might be used to counter any threat from a bombing that might occur (I would ask that the reader turn their attention back to the Denge Sound Mirrors that were mentioned in chapter one of this book. The same principle applies. The sound mirrors reflect the sound produced by an aircraft, and a radar station picks up the radio waves that have been reflected by a passing aircraft). In 1934, a large scale RAF exercise had taken place, with bombers aiming for London and fighter aircraft sent to stop them. The fighter aircraft had the Observer Corps to guide them to the targets. The result was that over seventy per cent of bombers managed to get through. This event led Winston Churchill to announce that the Royal Navy could not protect Britain from the threat of enemy attack (from the air) for the first time in history. In 1935, an experiment was conducted by Robert Watson-Watt, with others, which demonstrated that radio signals from a transmitter bounced off an aircraft travelling overhead. An experimental seaplane station was based at Felixstowe, and the aircraft was tracked for the first time. On 17 June 1935, radar was born in this country. This new technology was called pulse radar, whereby the time taken for the radio signal to return to the radar station could be measured. Radio waves travel at a known speed, the speed of light, which is 186,000 miles (300,000 kilometres) per second. For example, if the signal from the aircraft bounces back to the station 1/1000 of a second, then the plane must have a round trip of 186 miles, which makes the plane 93 miles away. The next stage in the process was the taking over RAF Bawdsey in Suffolk, England. Radio antennas were built at the site, with 240 ft (m) high wooden receiver towers and a 360 ft (m) high steel transmitter tower. This place was the RAF's secret radar base, where experiments were undertaken to perfect the techniques that would work so well in the forthcoming conflict. In 1940, the cavity magnetron was invented by the British. This equipment enabled powerful short wave radio signals to be emitted. This equipment could be fitted to ships and aircraft to enable easier tracking whilst the ship or plane was in motion. The government realised the strategic importance of radar and started building radar stations all along the south coast of Britain. The radar stations had antennas that were 360 feet (110 metres) tall. This equipment was called the Chain Home system. There are Transmitter Towers and other Listed buildings, one of which (Listing No: 1403955) is based at Swingate, Dover. There is a film that the reader may like to see called "Castles in the Sky", made in 2014. This film discusses the birth of radar in this country. It should be remembered that this was the first time in history that radar stations were built for defensive measures. This series of radar stations proved vital in the Battle of Britain (10 July till 13 October 1940), which was the only battle fought entirely in the air. The radar stations were also used for spotting V2 Rockets (more in another volume). The system set up to relay the information was called the "Dowding System". This system was named after Air Chief Marshall Sir Hugh Dowding (1882-1970), who realised that information had to be relayed quickly to be effective against enemy attacks. Private telephone lines were installed at all radar stations, and the information was transferred to Fighter Command, which could organise aircraft against enemy attacks. The plotting of enemy aircraft by the members of the WAAF (Women's Auxiliary Air Force, an organisation that carried out sterling work, though largely still unknown). This action was shown in the film "The Battle of Britain", made in 1969 with a stellar class of British actors. This now completes the twentieth-century application of radar.

In the second half of this chapter, I would like to discuss the uses of radar for grander purposes. In other Bits and Bobs books, I will mention both nature and medicine subjects. I want to start that discussion here. The first use of radar that I would like to discuss is the MRI (Magnetic Resonance Imaging) scanner. This technology uses the same principles of pulsed radiofrequency measurements. The original term was NMR (Nuclear Magnetic Resonance), but this was dropped as nuclear was considered a negative term (nuclear radiation, nuclear power plants). In brief, in the beginning, strong magnetic fields were used on particular objects containing water, and the amount of energy absorbed and reflected could be calculated. When an MRI scanner is used, you are placed inside a large magnet. A radiofrequency current is then emitted and reacts with the water inside the body's cells. An image is produced, and this image shows the bony structure of the body and soft tissues (such as the brain, liver, womb). This technology has helped detect disease, show how effective medication is working, and diagnose heart issues. This technology is undoubtedly one of the most significant advances in medical history (unless you count vaccines, though I will leave that debate for another book!)

There is another use of radar technology. We are now facing an uncertain future on a global scale. In the following chapter, I will be discussing the climate challenge that we all face. One of the tools that have been used to detect deforestation in the natural world is the use of radar technology. Deforestation is simply cutting down trees so that land which was once woodland now has another use. I have mentioned in a previous series of books, Pythagoras in the Corner, that most of the land that has been lost has been turned over to farming, either a monoculture such as palm oil or beef farming. This act of ecological terrorism is having devastating consequences for our planet. Let us have a look at some figures. The forest area in the world is roughly thirty per cent of the planet. However, with the rate of deforestation at present, it is estimated that by the year 2100, there will be no forest left in the world. So many articles and websites are dedicated to this issue globally, and I would recommend the reader find out more information for themselves. There is an organisation (of many, I hope in the future) monitoring this very situation. There are satellites (which I will discuss in a later volume) surrounding this planet and taking pictures using radar technology. This technology is enabling scientists from around the world to use this information. It is always life-affirming when people come together under a cause that is beneficial to everybody.

I have only included a limited number of examples in this chapter. The reader may be interested in meteorology, astronomy, and space exploration; all use radar technology (the reader can find out more for themselves). I hope this chapter has proved interesting.

CHAPTER NINETEEN – THE LETTER S

The following letter that I wanted to discuss is S, and I have chosen steel as my subject. I believe that this compound of materials has changed our world. It is the single most crucial element in construction and engineering. This chapter will discover how steel has helped our built environment become the place that we know and love today. We will also consider how steel weapons and war have changed our world. I hope the reader now understands the importance of this material.

I want to start this chapter by introducing how steel is created by mixing two metals. The material steel is principally made up of iron ore and carbon elements, with small amounts of other materials. Most steel is created in a furnace, and the reader can find out more for themselves. The inventor Sir Henry Bessemer (1813-1898) found the Bessemer process for making steel. This process was the principal process for making steel for the next hundred years. The weakness of cast iron (the Tay Bridge disaster, 1879 is an example) made steel one of the critical components of the Second Industrial Revolution. The built environment changed when steel was found to have higher strength capabilities than other materials, such as wrought iron. The discovery of rolling steel beams into one long shape changed how we live our lives. The reader should understand a common point. The safest structure is a pyramid, as this has a large base and graduating sides to a point at the top. The base enables any movement in the ground below to be negated. However, as we know, the land is scarce, and therefore pyramids use space. As cities became more congested (the Industrial Revolution started this), slums started to appear as older housing was used and re-used by the poorest of the population. Therefore, buildings started to get higher (I have discussed this in another book PITC), and new technology was required. As we fast forward, the ultimate high rise building is known as the skyscraper. It is this type of structure that changed the cities that we created. The foundations of these very tall buildings are dug incredibly deep, as they have a large amount of weight that is to be supported by the foundations. Modern skyscrapers have concrete piles driven into the ground, from which a steel beam is placed vertically. There are several steel beams at various points, and horizontal beams will tie these together. This system is suitable up to forty floors high. Once a certain level is reached, other types of steel beam construction are used, such as tubular steel construction, which is more cost-efficient. This material was developed in 1963 by Fazlur Khan and J Rankine. The first building that used steel construction was the Home Insurance Building in Chicago, America. This building was built in 1885, though now sadly demolished. The Royal Insurance Building, located in Liverpool, was built between 1896 and 1903 and was the first building to use a steel frame in the United Kingdom. The building is now a hotel, the Aloft Hotel in Liverpool. The Gherkin (30 St Mary Axe), or Swiss Re building in London, is a curved skyscraper that uses glass and steel in construction. There are 35 kilometres of steel in the building with over 38 floors to climb. The building took two years to complete and cost 204 million pounds. It is 180 metres high and has 7,429 panes of glass. It was sold to a Brazilian billionaire Joseph Safra for £700 million. Before I leave this section, there is one building that I have to mention. The United States Air Force Academy Cadet Chapel. The building was completed in 1962 and is a tubular steel frame construction. The building has seventeen twin spires and tetrahedrons (triangular pyramids with four triangular sides, six straight edges and four corners). These spires are 23 metres (75 feet) high, and each weighs about five tons. Between the spires, the coloured glass fills the gaps. There are two structures that I would like to include before we finish with the plus side of steel as a material. In Gateshead, Tyne and Wear, an iconic sculpture called the "Angel of the North" by Anthony Gormley (born 1950). It is a steel sculpture of an angel, completed in 1998, that is sixty-six feet (twenty metres) high, with wings of fifty-four feet

(177 metres) wide. There is an exciting sculpture that I would like to mention. During the Irish Potato Famine (mentioned in the Introduction), the Choctaw Nation in Oklahoma, the United States, in 1847, donated $170 ($5000, five thousand dollars in today's currency) to the Irish population to assist their suffering. The Kindred Spirits sculpture consists of nine twenty-foot (over six-metre) high stainless steel feathers arranged in a circle. The sculpture in County Cork symbolises love and trust between the two nations. There is one more statue that I want to mention. The Genghis Khan Equestrian Statue is located in Mongolia. It is the highest equestrian statute globally and is made of stainless steel. The reader can find out more for themselves.

Another side to steel has led to whole civilisations being overrun by their enemies. The making of steel was for one purpose only: the sword. Whilst these have a ceremonial function today, there is a long and fascinating history of swords (the reader can find out for themselves). Swords such as the spatha in Ancient Rome and the longsword in Medieval Britain have been written into history. There is one sword that deserves a special mention. The sword used by samurai warriors in Japan is called the katana. These swords were made between one hundred and fifty to four hundred years ago for the noble elite of Japanese society, the samurai. Therefore, the most expensive and original sword comes from Tama-hagne steel. Modern replicas now use steel alloys. Manufacturing a samurai sword involves many hours of hammering and folding the steel back onto itself. By folding the steel many times, for example, fourteen times of folding produces sixteen thousand layers of steel. This sword deserves its place in history because of the way the Japanese used this weapon to defeat their enemies. The most famous of these was the defeat of the Mongols led by Genghis Khan (the reader should find out more for themselves).

The last section in the chapter deals with the joining of two steel plates together. There are two methods, one method is called riveting, and the other method is called welding. Riveting is the method of putting permanent mechanical fasteners into pieces of steel to hold them together. The riveting process involves joining two plates of metal, pushing through the hot rivet, and smashing the end of the rivet to hold the metal in place. The riveting process is very labour expensive and requires skilled artisans. Examples of riveted structures

range from the Eiffel Tower to the RMS Titanic. The Belfast shipyard that built
the RMS Titanic had over thirty thousand people on the payroll. Regarding the
Titanic, there were a total of three million rivets; however, some of these were
poorly made. The rivets popped open on impact with the iceberg, allowing seawater to flood the vessel. The other process is called welding. This process also
uses heat; however, the work is less skilled and less labour intensive. Intense
heat is applied to make the two joined pieces of metal molten. A filler material
is added, and the metals mix to form one piece of metal when cooled. Using the
welding technique, more ships could be produced at a far lower price than before. In a previous chapter, we mentioned Liberty ships. These ships were built
in America and were cheap to manufacture. Due to the sinking of merchant's
vessels, these ships were required by the United Kingdom in a conflict that
came to be known as the Battle of the Atlantic (readers can find out more for
themselves). This war was the Allies' blockade of goods and food in Germany.
The German navy responded in kind, with submarines (U-boats), ships and
aircraft against Allied merchant shipping vessels. This warfare was one of the
longest continuous theatres of warfare, lasting throughout the Second World
War. The British government ordered liberty ships from the American allies,
and over 2,710 ships were completed. This workload equates to three ships
being made every two days (a riveted ship could take several months to build).
However, these ships were only designed to have a five-year life span, whereas
a riveted ship would last a hundred years. Therefore, during the later stages of
the Second World War, more welded Liberty ships were required as ships were
being sunk faster by the enemy. The only disadvantage was the poor welding,
which meant that in the early days of the Liberty ships, over fifteen hundred
ships were lost to brittle fractures (mainly due to the low temperature of the
Atlantic). The Liberty ship programme was one of the contributing factors to
the Allies winning the war (alongside the weapons, jeeps, ships and aircraft
that also contributed, these will be discussed in another book).

In this chapter, I hope I have shown the importance of steel to both the
twentieth and the twenty-first centuries. We will continue to use this incredibly
versatile material for many years.

CHAPTER TWENTY – THE LETTER T

There are many engineering terms for the letter T. The word chosen is the tunnel. There is not enough time and space within this book to include every tunnel, as there are thousands of examples across the globe. The author has tried to include some of the tunnels that have either been famous (or infamous) or been used for an unusual purpose. I will not try and cover the history of tunnels, as this is far too complex and long for this book.

The first point to note is that all tunnels are horizontal. The second point to note is that they are both long and narrow. The length of a tunnel can be up to several miles, which can create serious issues, and accidents can occur (we will discuss these later). Let us now look at how tunnels are used every day. Some tunnels have been constructed to carry water or waste, which form an essential function. A tunnel can transport people and goods, such as a road tunnel or canal tunnel. Many trains use tunnels; indeed, tunnels can be seen on all major rail routes, such as the Great Western Railway. Not all tunnels are underground; there are examples such as wind tunnels, which will be discussed later. There is another form of tunnel that I will call the War Tunnels, and these will also be included in this chapter.

The first example I would like to mention is the Crypta Neapolitana (also known as the Pausilippo Tunnel). This tunnel was constructed in 37 B.C., during the time of the Roman Empire. The tunnel is based in the city of Naples, Italy. The tunnel was created to link the city of Naples with the city of Pozzuoli (now part of Naples). During the tunnel's creation, there was a marshland that separated the two cities. An overground road would be an impossible option; therefore, a tunnel was created. The tunnel was 4800 feet long, 25 feet wide and 30 feet high. It is unknown how many enslaved people were used to create this engineering masterpiece, though it is believed to be in the thousands (we will return to this theme in other books as well).

In chapter two of this book, I mentioned the Drogden Tunnel from Peberholm to the Danish island of Amager. This tunnel is part of the Oresund Bridge, a five-mile bridge which connects Copenhagen in Denmark to Malmo in Sweden. The overground part of the bridge is from Malmo in Sweden to an artificial island called Peberholm in the Oresund Strait. The tunnel is over four thousand metres long (2.5 miles) and has five tubes in the tunnel. Two tubes are for railways, two tubes are for cars, and the third tube is for emergency vehicles. I have included this bridge and tunnel as it forms the backdrop to the crime series The Bridge, which started in the United Kingdom in 2012. Of course, I would recommend the reader enjoy this fantastic series and the bridge itself.

In chapter three, I mentioned a fantastic engineer called Joseph Bazalgette (1819–91), who engineered the first sewerage system in London. This sewerage system is based under a reclaimed part of the River Thames called the Embankment.

During this time in London, there were several disease outbreaks, including cholera. Many of these diseases can be traced to unclean water, with raw sewerage as part of the water! Therefore, a new system had to be constructed, and it fell to Joseph Bazalgette to design the system. To determine the diameter of the pipe required, several calculations had to take place. There was a statistical calculation looking at the population growth expected in the city and a calculation of the effluent to be produced. Once a route had been established, Joseph Bazalgette had the brilliant idea of doubling the diameter to account for population growth. This doubling meant a far more giant tunnel; however, this pipe proved helpful for the next hundred years plus. However, there is now double the number of population living in London that was ever envisaged when the tunnel was first constructed. Therefore a new system called the Thames Tideway Tunnel, and the Lee Tunnel have to be constructed. The Lee Tunnel would be seven metres in diameter and travel 6.5 kilometres between

Abbey Mills Pumping Station to the Beckton wastewater treatment plant. The tunnel would cost around £635 million to construct and capture 16 million cubic tons of sewage from London. This tunnel is one of the deepest tunnels ever bored in London. The tunnel is bored from 75 metres below the surface at Abbey Mills to 79 metres deep at Beckton. As part of this project, Beckton wastewater treatment plant would be upgraded to cope with the increased demand and be able to handle the anticipated sixty per cent increase. The Thames Tideway tunnel will be 25 kilometres long and will follow the course of the River Thames from Acton in west London to Abbey Mills pumping station in the east of London. The tunnel will be dug to a maximum depth of over 60 metres deep and over seven metres in diameter. The Thames Tideway Tunnel will connect with the Lee Tunnel, mentioned earlier. The project's cost is deemed to be in the region of over £4 billion, and the cost will be met by an increase in Thames Water bills by the public.

Another example is the use of a tunnel for hydro-electric power. There are two examples that I would like to mention, one in the United Kingdom and the second example in America. In the United Kingdom, the first major pumped storage hydro-electric power station is the Ffestiniog Power Station in Wales. It was completed in 1963 and provides 360 MW of electricity to the National Grid. The system works between two reservoirs. When power is required, the top reservoir called Llyn Stwlan starts to empty. The water is forced down two 200 metre shafts that deal with a pressure of 27 cubic metres of water a second. The water is then forced into four concrete tunnels that feed the water through the turbines' valves. The water then proceeds onto another reservoir called Tan-y-Grisiau. The water is then pumped back up to the top reservoir (Llyn Stwlan), and the cycle starts again. The second example concerns a power station based at Niagra Falls. In 1895 at Niagara Falls, the world's first large-scale hydro-electric power plant was installed on the border between America and Canada. This plant was called the Edward Dean Adams Power Station and was the largest of its kind globally at the time. The Adams power station was opened amid great publicity and was said to start the 'electrification of the world'. The plans for the site included building an enormous tunnel (21 feet high and 18 feet wide) for the water to flow through that would power the generators. The tunnel was around 6,700 feet long, and 600,000 tons of material were removed. The excavated material was used to extend the shoreline of the Niagara River. Construction materials for the site consisted of 16 million bricks, 19 million feet of timber, 60 million cubic yards of stone and 26,000 cubic yards of sand. Over 67,000 cubic yards of cement were poured. The tunnel took about three years to build and cost 28 lives. In 1896, the City of Buffalo received the first power from the Niagara Falls power station, relayed by 25 metres of electric transmission lines.

Before I finish this section, there are some famous tunnels that the reader may wish to discover on their own. The first tunnel is the Woodhead Tunnel; a 3.5-mile tunnel dug for Manchester-Sheffield Railway in 1845. The next tunnel is the Hoosac Tunnel in Massachusetts for the Vermont Railway network. This tunnel was the first large tunnel in the United States. It was built over 20 years (finished in 1876) and was 4.75 miles long. This tunnel was one of the first tunnels that used the power of the dynamite to cut through the rocks and compressed-air systems. The railroad tunnel was nicknamed the "Bloody Pit" for taking the lives of hundreds of people during its construction. The next tunnel is the Thirlmere Aqueduct, which is incorrectly thought to be the world's longest tunnel (the tunnel, however, is not continuous). However, the Thirlmere Aqueduct is the longest in Britain and has no pumps along its route into Manchester. The system is gravity fed, and the water drops twenty inches per mile along its route. The Mont Cenis Tunnel was the first of many tunnels that now crisscross the Alps. The giant undersea tunnel is the Seikan Tunnel in Japan. The longest railway tunnel is Gotthard Base Tunnel in Switzerland. Finally, Norway's longest tunnel intended for car transport is Lærdal Tunnel.

I want to mention a set of tunnels now (they can be found under Stalag Luft 3 in Google). During the Second World War, many captured Allied soldiers were held in POW (Prisoner Of War) camps. One of these camps was known as Stalag Luft III, about 100 miles from Berlin, in Poland's occupied territory. The site was chosen as the sandy soil in the area made tunnelling difficult for any proposed escape. However, there were two escapes, both of which were tunnels, made into films! The first escape tunnel was constructed in the open, only feet from the perimeter fence. The Germans were unaware, as a piece of gymnastic equipment covered the hole every day. The film is known as The Wooden Horse (1950) and depicts a true-life attempt. The main actors in the film also had exciting careers. David Thomlinson (1917-2000) was an actor in many films, including starring as George Banks in Mary Poppins (1964, one of the author's favourite films). Another actor, Leo Genn (1905-78), was a barrister in real life. The barrister was an assistant prosecutor at the war crimes trials for the atrocities committed at Belsen concentration camp. There was another breakout at the same camp; tunnels were dug underneath the accommodation huts. Three tunnels were dug, nicknamed Tom, Dick and Harry. The film "The Great Escape" (1963) shows many true-life events. The actual escape happened in March 1944 and seventy-six prisoners escaped, and only three made their way to safety. Fifty of the escapees were shot. I will let the reader find out more for themselves.

The last structure that I would like to discuss is not a tunnel but a sizeable artificial hole. It is known as a cistern (similar to the domestic variety). It is called the Basilica Cistern in Istanbul, Turkey. The cistern was built in the sixth century A.D., during the time of the Byzantine Empire. Let us consider the dimensions. The cistern is nine metres (thirty feet) high and is supported by 336 marble columns. The area of the cistern is 9,800 square metres (138 metres (453 feet) by 65 metres (213 feet). The cistern can hold eighty thousand cubic metres (280,000 cubic feet) of water. The cistern continues to provide water to Topkapi Palace to the present day. I have included this structure as it is featured in the James Bond film From Russia with Love (1963). I will do another volume (called Bits and Bobs of James Bond) in the future.

One of the unique parts of tunnel engineering is the engineers themselves. We have already mentioned Joseph Bazalgette, and I would like to introduce a famous father and son team for their engineering work. The engineers are Sir Marc Isambard Brunel (1769-1849) and his son, Isambard Kingdom Brunel (1806-59). The most outstanding achievement of Sir Marc Brunel was the building of the Thames Tunnel between Rotherhithe and Wapping, under the river Thames. This tunnel was started in 1825 and completed in 1843. This was a tunnel that was thought impossible due to the nature of the soil and the risk of flooding. The risk of collapse of the tunnel was overcome by building a shield that protected the miners. However, during the construction of the Thames Tunnel, the water that leaked into the tunnel contained sewage, which made many of the miners ill, including Isambard Kingdom Brunel. The

Thames Tunnel also cost the lives of six men who died during the flooding of the tunnel. The tunnel was converted from pedestrian traffic to railways and is in use today as part of the East London Line. Let us now consider his son, Isambard Kingdom Brunel. This man can be called one of Britain's most excellent engineers. His engineering work building bridges (the Clifton Suspension Bridge, amongst others), ships and railways (the Great Western Railway, amongst others) has never been equalled. Isambard Brunel also created, at the time, the world's longest railway tunnel called the Box Tunnel. This tunnel is located between Swindon and Bath, on the Great Western Railway line. The tunnel is less than two miles long, which seems small compared to the present day. However, unlike modern-day techniques, the tunnels were lit by candles and dynamite was used whilst men were still in the tunnel, working nearby. The flooding and the blasting took the lives of around one hundred navvies during its construction. Parts of the Tunnel are now Grade II Listed.

Before we leave this section on people, there are two very eccentric gentlemen that I would like to introduce. The first gentleman is the 5th Duke of Portland, William John Cavendish Cavendish-Scott-Bentinck (1800-79), who lived at Welbeck Abbey, Nottinghamshire. He was a known eccentric who did not enjoy the company of other people and never invited people around to his house. He constructed a vast underground network of tunnels and underground chambers and employed hundreds of local men on these enormous construction projects. The tunnels were estimated to be over fifteen miles long, and large underground chambers were constructed. One of these chambers was large enough to be a ballroom, over 49 metres long by 19 metres wide. The ballroom could be accessed through a hydraulic lift that could carry up to twenty people. It was never used as a ballroom, as the Duke never had guests. There was also a 76-metre vast library and a billiards room. The Duke died childless in 1879. The next gentleman is Mr Joseph Williamson (1769-1840), a businessman, philanthropist and property owner. His houses on Mason Street were "unorthodox" and were built without any building plans. Williamson is also known as the Mad Mole for the Williamson Tunnels constructed around the Edge Hill area of Liverpool. The tunnels ranged in depth from ten feet to around fifty feet and went on for several miles. The author believes, as do others, that his philanthropy caused him to employ workers digging out tunnels.

It should be remembered that at this time, there was a sizeable unemployed workforce, straight after the Napoleonic Wars (the reader can find out more for themselves). Men could be employed and receive a wage whilst still retaining their self-respect. That act of philanthropy alone earns the author's respect. The current Doctor Who series features Joseph Williamson as the Mad Mole (episode aired on 31st October 2021). That is a fitting tribute. On tunnels and television, a series called "The Peaky Blinders" should be mentioned. The main protagonist, Thomas Shelby, was employed as a tunneller during World War One. It may be interesting for the reader to learn about the Tunnelling Regiments of World War One. They will not be disappointed.

Wind Tunnels are the last part of this section on tunnels. A wind tunnel can be above ground and is simply a tube with air flowing inside. The air moving inside the tunnel replicates an aeroplane in flight. In the first chapter, we discussed aerodynamics and aviation history. As aircraft speed increased, so did the forces that reacted with the aircraft. Therefore, more extensive and enormous tunnels were needed to be constructed. During my time as a surveyor, I worked in the town of Farnborough, in Hampshire. This town has many connections with aviation. Before the computer age, wind tunnels were used at Farnborough from the 1930s. This place was the birth of modern aviation, and the history of the Royal Aircraft Establishment makes fascinating reading. It should be mentioned that NASA has wind tunnels too, and these are used for spacecraft on interstellar voyages. I would urge the reader to find out more for themselves.

I hope this chapter has been of interest to the reader and is not too dull. There was so much to discuss that I could have filled volumes of books about tunnels (though they would have been difficult to market to people outside of the construction and engineering profession). I hope this chapter leads people to make exciting discoveries.

CHAPTER TWENTY-ONE – THE LETTER U

D ue to climate change, we are dealing with unprecedented weather conditions in recent times. Events such as drought, flood, fire and storms are becoming regular rather than rare events, and these are some of the terrifying natural events in extreme circumstances. Therefore, we have to consider our greenhouse emissions that are altering the weather on our planet.

For the letter U, I have chosen a measure called the U-Value. When considering the greenhouse emissions that we discussed earlier, we must look at how buildings contribute to this problem. It is estimated that construction activities and buildings are responsible for nearly forty per cent of all greenhouse emissions, and the emissions from energy used to heat, cool, and light buildings account for nearly thirty per cent. Therefore, we should consider how buildings are heated (cooling and lighting will be ignored for now).

The U-Value measures the amount of heat loss through a material or part of a building such as a floor, roof or wall. It is measured in watts per square metre, per degree Kelvin, W/m2K. In order to understand this measure, we need to go backwards before we leap forward, and two measures require explanation. The Lambda Value measures heat loss and the R-Value measures resistance to heat loss.

The Lambda Value (λ) is the rate at which heat is transferred through a material (thermal conductivity). The Lambda value is measured in Watts per square metre of area for a temperature of one Kelvin per metre thickness (W/mK). The most practical example is considering the rays of sunlight warming the Earth, measured in Watts per square metre. We need to measure the amount of heat loss in Kelvins (mentioned earlier) and the thickness of the material used. Thermal conductivity is Fourier's Law, named after a French physicist, Joseph Fourier (1768-1830). Fourier discovered that certain materials allow heat to pass through them easier than others. (He was also concerned about the greenhouse effect, though that term was not used then). So copper allows heat to pass through easier than steel, for example. A practical example is copper piping used in houses to carry heated water around.

The R-Value stands for resistance to heat loss. As mentioned earlier, the R-Value is the inverse of the Lambda Value; therefore, it is $1/(\lambda)$. R values are expressed in m2K/W. The area of the object is known as m2, as mentioned earlier. K stands for Kelvin (a measurement of temperature), and W stands for Watts; in this case, the heat is transferred through a building. All materials allow heat to transfer through them. Therefore, the R-value measures the performance of the material used when considering the resistance to heat loss. It is pretty simple to understand R Values, as a higher number stands for a better type of material. If a material has an R rating of 15, it functions better than a material with an R rating of Nine. Let us consider an example. If we use a standard house brick, we can place a thermometer on one brick side. By considering the temperature on both sides of the brick, we can measure how much heat is conducted through the brick. This temperature will also show how resistant the brick is to changes in temperature. The same can be seen in a cavity wall, where the heat transferred from the internal wall of the house to the external wall can be measured. So, what are the R Values of a cavity wall? We have an outer brick wall, insulation, and a concrete block inner wall for a cavity wall (ignoring the plaster). The R-value of a brick, say, is 0.8, the insulation (which can be a different number of materials, let us say solid foam with a foil on one side) has an R-value of 7.20 and the concrete block inner wall, let's say 0.80. So the total resistance to heat of this wall is 8.80 m2K/W.

The third and most important measure is the U-Value. As mentioned, U values measure heat loss through a material or a part of the building such as a floor, roof or wall. So, what is a U Value? It is watts per square metre, per degree Kelvin, W/m2K. This value is the inverse or opposite of the R-Value, as we measure the flow of heat rather than the resistance to heat loss, as mentioned above. A U Value is simply the inverse of all the R Values (mentioned earlier) added together = 1/r. S So, from the equation we have, from the cavity wall mentioned earlier is 1/r1(brick)+1/r2(insulation)+1/r3(concrete block) = 1/r1+r2+r3. So why are U values more important than Lambda values or R Values? U values, as mentioned, measure the flow of heat through an insulating material; therefore, the lower the U value, the greater the insulation of the product and less energy is used to heat the building.

In order to further our understanding, I would like to show how a building can be improved. Most modern houses, and I will only be dealing with domestic dwellings at this point have cavity wall insulation and a damp proof course (DPC). When we consider the house, we should look at all the elements of the house and the operational parts of the house. When works are undertaken to improve a dwelling or premises, a term is used called renovation. This term covers all manner of building operations so I will use this term from now. To improve the energy efficiency of a house, we could increase the insulation of the loft, external walls and even the roof itself. Concerning the windows, you would replace these with energy-efficient double-glazing in a domestic dwelling. Another example would be to improve the draught-proofing so that less heat is lost through the doors and windows. Another domestic dwelling improvement would be to install an energy-efficient boiler; this would save both the occupier money and help the environment. There are other measures, such as solar panels, which reduce the amount of electricity used by the building. There are air conditioning systems in some dwellings that help cool and even filter the surrounding air. The refrigerant used should be current and within guidelines, and a professional body should check the system.

There are two examples that I would like to discuss. Both of these examples concern famous buildings. The first example is the renovation of an old building, and the second example is a brand-new building. The first example that I would like to discuss is one of the most famous buildings globally, the Empire State Building in New York. This skyscraper was constructed in 1931 and, from 1931 to 1970, was the tallest building in the world. However, with such an old building, there are massive operating costs. In 2010, the Empire State Building underwent a massive renovation, with energy efficiency works costing $120 million. The main aim of this project was to reduce the amount of energy used to heat the building. The windows in the building were old; however, the cost of a complete replacement proved too expensive. The solution was to reuse the old glazing. An E-film was placed between existing panes of glass, which allowed light into the building but blocked heat gain through the windows. As explained above, the R-Value increased from 2.0 to 7.0. This meant that air conditioning costs were reduced, massive saving in energy bills. Other energy-efficient measures included demand-led lighting by tenants and better ventilation controls. The total saving is over $4 million per year.

The following example is the world's tallest skyscraper, the Burj Khalifa in Dubai, United Arab Emirates. This building has a total height of 829.8 metres (2,722 feet), completed in 2009. The building uses solar panels to heat over a hundred and forty thousand litres of water every day. The total energy saving is 690 MWh every year. The building is incredible to look at, and solar panels ensure that the building's electricity needs are met for the future.

I hope this small chapter gives you a taste of the complicated and complex energy efficiency problems and how these can be overcome.

CHAPTER TWENTY-TWO – THE LETTER V

There was only one option that I wanted to discuss for the letter V, and that was V-weapons. These were ballistic weapons called Vergeltungswaffen, or retaliatory weapons. These were the fabled secret weapons that Hitler boasted about, the weapons that would win the war for Nazi Germany. This boast was due to the failure of the Luftwaffe, the German Air Force to win the sky over Britain (the Battle of Britain, the reader can find out more for themselves). The V-Weapons were long-range ballistic weapons used for terror bombing (indiscriminate bombing) or aerial bombardment of Allied cities. There were 3 V-Weapons, the V1, V2 and the V3. The V weapons were built at Peenemunde, a remote island off the Baltic. Here, the Nazis had assembled a group of scientists and a workforce who worked under the greatest of secrecy. They were called the Peenemunde Army Research centre (the reader can find out more for themselves). Under great secrecy, the Polish underground movement sent back information about the Army base at Peenemunde. In their report, they mentioned the unusual concrete ramps erected at the site required for the missile's trajectory. During the Second World War, at RAF Medmenham, the RAF developed a Photographic Intelligence Unit, which examined photographs taken over enemy territory. Modified Spitfires took pictures over enemy territory. They had no weapons and were a blue/grey or silver colour to blend in with the sky. The Spitfires flew at over 30,000 feet to take their photographs. By the end of the War, there were 36 million negatives and pictures for the staff to pore over. (A small aside for the reader. I am a member of the National Trust and once visited Hughenden, the home of the Victorian statesman and Prime Minister, Benjamin Disraeli (1804-81). At this house, during the War, there was a map-making facility used by RAF Bomber Command to make maps of enemy territory. In the ice house at Hughenden, they had the perfect darkroom to develop the photographs taken. I have seen the pictures taken of Peenemunde here). In August 1943, Operation Crossbow was given the go-ahead by senior Civil Service Staff to bomb Peenemunde and other sites (see later). Before I begin this section in detail, I would like to relate how I became interested in this subject. I was working as a surveyor for the London Borough of Hounslow, in the Chiswick area, when I came across a plaque in Staveley street. The plaque was erected as this earmarked the site of the first intercontinental ballistic missile attack in the United Kingdom (the first V2 missile attack). So started my interest in this topic. However, before we discuss these weapons, we should understand the historical context of these weapons. I will also discuss the technology (in simple detail) and the people behind these weapons. Let us now examine these weapons in this order.

The V1 weapon was built in 1944 by the Gerhard Fieseler Werke company in Germany. An engineer named Robert Lusser (1899-1969) designed the VI weapon. (There are 67 defunct German aircraft manufacturing companies, including Heinkel, Junkers, Messerschmitt and Zeppelin-Staaken, the reader can find out more for themselves). The Fieseler company was named after Gerhard Fieseler (1896-1987, an aerobatics champion and WW1 flying ace). The V1 had a wingspan of plywood and measured over seventeen feet (over five metres). The aircraft's body measured just over twenty-seven feet (over eight metres). The VI missile weighed 4,750 lbs (over two thousand kilograms). The V1 was a pulse-jet cruise missile, the first of its kind globally. I want to break down this description in detail. A pulse-jet engine consists of a body containing a mixing chamber for the fuel and air, flaps and valves and a spark plug. In simple terms, the compressed gas inside the chamber would be ignited, and the exhaust flowing out the back creates the thrust that powers the missile. The V1 was nicknamed the "doodlebug" or the "buzzbomb", as that was the engine's noise as it flew past the onlooker. A cruise missile is a guided missile which travels towards its intended target at a constant speed (the V1 travelled at 400mph). The V1 had a maximum flying distance of 200 miles, but the weather could decrease this. A pre-set magnetic compass and gyroscopic auto-pilot determined and maintained its course. A small propeller at the front of the weapon registered the distance covered. The guidance system cut the power to the engine at a pre-set distance, and the V1 went into a steep dive. The V1 carried one ton of high explosives and travelled at a maximum of 400 mph. The warhead consisted of 850 kg of Amatol, 52A+ high-grade blast-effective explosive with three fuses. An electrical fuse could be triggered by a nose or belly impact. Another fuse was a slow-acting mechanical fuse allowing deeper penetration into the ground, regardless of the altitude. The third fuse was a delayed-action fuse, set to go off two hours after launch. The purpose of the third fuse was to avoid the risk of this secret weapon being examined by the British. It was too short to be any booby trap but instead meant to destroy the weapon if a soft landing had not triggered the impact fuses. These fusing systems were very reliable, and almost no dud V-1s recovered. The last point in this section is the terrible loss of life caused by this particular weapon. The first weapon was launched on 13th June 1944 and struck a railway bridge at Grove Street, Bethnal Green. Six people were killed in that initial attack. Between 1944-and 1945, 2419 V1 rockets were fired from enemy-held positions in France (Pas-de-Calais, France, was overrun by Germany during the War). There were over six thousand people killed and over seventeen thousand casualties. In Chapter One, I asked the question about the V flying formation. There was an unfortunate incident with the 102 Bomber squadron. A bomber dropped a bomb on a bomber flying beneath them, on a mission to destroy a V-weapon facility in France (the reader can find out more for themselves). Since that incident, planes always fly in a V formation to stop the incident from occurring again.

The second V weapon, the V2 Rocket, was a far more dangerous weapon. This Rocket was the world's first long-range guided missile. The V-2 was 14 metres (47 feet) long, and weighed 12,700–13,200 kg (28,000–29,000 pounds) at launching. The alcohol and liquid oxygen mixture developed about 60,000 pounds of thrust. The payload was about 1,000 kg (2,200 pounds) of Amatol, as per the V1 mentioned above. The range of the V2 Rocket was about 320 km (200 miles), and the peak altitude usually reached was roughly 80 km (50 miles). The Rocket travelled at supersonic speed (speed greater than the speed of sound (Mach 1) at 343 m/s (metres per second) or 767 mph (miles per hour). The speed of the V2 Rocket was 5760 kilometres per hour (1600 metres per second) or 3,580 mph (miles per hour). This speed would be known as Mach 5 speed. There was simply no defence against such a weapon during World War Two. The V2 Rocket was developed simultaneously with the V1 Rocket at the Army base at Peenemunde (as mentioned earlier). The engineering genius was Wernher von Braun (1912-77) and his team at Peneemunde. We will come back to this person later. There were over three thousand rockets produced, though not all at Peenemunde. The Germans moved the site due to the Allied air raids that had decimated the compound. As mentioned, there was no defence against this weapon. The Allies decided that the Germans would need convincing that the centre of operations had moved from central London to Dulwich, east of London. This was achieved by using, amongst other factors, double agents; I will discuss these later in the chapter. This deception meant that many V2 missiles were aimed away from Central London (it has been commented that many of the missiles fell short, though this was unknown).

It is now time to discuss what was happening in Europe in the Second World War. In short, the German Army was starting to lose the War. On 6th June 1944, Operation Overlord commenced. This event was the Allied invasion of Normandy and became known as the D-Day landings (thanks to the work of Juan Garcia, amongst others, the Germans were convinced that the landings would be in Calais, not Normandy. This double-cross saved the lives of hundreds, possibly thousands of servicemen). This event started the long road to the defeat of Nazi Germany. As Allied bombing raids were getting better results, Albert Speer (the Nazi architect and head of Armaments 1905-81) ordered all armament factories to be built underground. We should never forget that all these factories throughout Germany, France, Poland and other conquered territories had to be supplied by enslaved labour. It is estimated that during World War Two, over fifteen million forced labourers were working for the Germans from occupied territories (see Forced labour under German rule during World War II). It has been estimated that at Mittlewerk, an underground V2 factory, of the 60,000 slave labourers who worked there, only 20,000 survived. In the author's view, one of the most famous underground factories was La Coupole in northern France. This factory was built between 1943-and 44, with over seven kilometres of tunnels underground. The structure's dome is seventy-two metres in diameter, and 5.5 metres thick, created from over fifty thousand tons of concrete. This type of installation would not be constructed unnoticed, and part of Operation Crossbow was to destroy the facility. It was eventually discarded. However, I will leave the reader with one chilling thought. When La Coupole was to become operational, the mandate was to fire forty to fifty V2 rockets a day targeting London. Luckily, this never happened. However, about 1,100 V2s were fired at Britain before the advancing Allies overran their launch sites. In total, they killed or wounded about 115,000 people. Antwerp, a vital port for the Allies, was devastated by attacks from V weapons, but, in general, they were used on civilian targets only. The V2 Rocket caused nearly three thousand civilian deaths, with over six thousand people injured or maimed.

At this point, some people should be mentioned who greatly assisted the Allied war effort against the V Weapons programme. The first person that I would like to introduce is Reginald Jones (1911-97), an MI6 Officer who was Assistant Director of Intelligence (Science), based at the Royal Aircraft Establishment at Farnborough, Hampshire. There are two measures that Jones introduced that helped win the Second World War. The German bombers were using a " Knickebein " system, which allowed bombers to navigate by night to reach their intended target. This system used radio signals that enabled the Germans to follow a specific route. By intercepting the radio signals and bending the beams, bombs fell uselessly on the English countryside. This radio war became known as the "Battle of the Beams" (the reader can find out more for themselves). The second tactic was to fool radar signals using tin foil (known as chaff). Jones also served as an expert on the V2 Rocket system and was instrumental in the Double Cross system, which used double agents. The following person is a double agent by the name of Eddie Chapman (1914-97). This agent, who the Germans trusted, convinced them that the targets of the V2 rockets were wrong. Eddie Chapman may have saved hundreds of civilian lives by giving the wrong map coordinates. The following person is Wulf Schmidt (1911-1992), a Danish double agent (Tate) who helped change the War. Tate was one of the longest-running double agents of the Second World War. His information regarding a fake USAG (First United States Army Group, never existed), targets for V2 attacks (way off the mark), and a minefield for U-Boats off Ireland (never existed), the Germans awarded Tate the Iron Cross, First and Second Class! This story has to be one of the must-reads for the reader. The last person I would like to mention is Jeannie Rousseau (1919-2017), a French Intelligence Agent during the War. Her intelligence gathering regarding Peenemunde, and the information sent back to the Allies, saved the lives of thousands of civilians during the War.

The last weapon that I want to discuss is the V3 weapon. It was, in fact, a giant cannon, being 130 metres (430 feet) long. The V3 weapon was projected to fire about six hundred shells an hour. The missile was fired using a form of multiple propelling charges that would keep the projectile moving along the barrel. The estimated target range of the Rocket was about one hundred miles. This range means that the gun would easily have been able to hit London and other European cities. However, before the weapon finished its initial tests at a base outside Peenemunde, the site was bombed by Allied bombers. A new site was chosen, with the gun buried in a tunnel inside the cliff, with supporting tunnels housing the ammunition. The site was called Mimoyecques, near the Pas-de-Calais. The weapon was permanently aimed at London. The site was put out of action when the RAF 617 Squadron (the famous Dambusters, see earlier) used specially adapted missiles known as "tallboys" (more in another volume). However, a modified form of the V3 weapon (only fifty metres long) did go into action at the site. The target was Luxembourg, with ten people dead and thirty-five wounded. Many of the V3 sites were abandoned, with the guns dismantled. When they found the area, some of the weapons were found by American forces. The equipment (gun tubes, ammunition and spare parts) was returned to the United States. There is an epitaph to the V3 weapon. Superguns have continually been developed after the Second World War. I will, in another volume, discuss the HARP gun amongst other projects that have been undertaken.

It is perhaps fortunate that all of the V-Weapons programmes came to nothing. Despite the high number of dead and wounded people caused by these weapons, they did not alter the course of the War. The fact was that the weapons were developed too late to be effective. The German Army ran out of time, and the Allied bombardment meant these weapons never met their full potential. For that reason alone, we should be grateful.

I want to finish this chapter on a more positive note. In brief, after the Second World War, many scientists, including Werner von Braun, were taken to America. This operation was called Operation Paperclip, meaning a new sheet of paper was paperclipped over their war record, which the Americans and the Soviets ignored. There is still controversy over this move, as many believed they should have been accountable for their actions. However, some of these scientists went to NASA (and the Soviet equivalent) and assisted with the space programme. Eventually, Man went to the Moon, which will be discussed in another volume.

There is one more film that I have to mention now. The film is called Dr Strangelove or: How I Learned to Stop Worrying and Love the Bomb. A 1964 film starring Peter Sellers (1925-80) as the lead character. This film is a black comedy about the fears of a nuclear conflict. I have mentioned this film as the lead character is an amalgamation of several people, including Werner von Braun (discussed earlier), Herman Kahn (Head of the RAND Corporation), John von Neumann (Manhattan Project) and Edward Teller (father of the H-Bomb). I would recommend the reader see this film!

CHAPTER TWENTY-THREE – THE LETTER W

There is one invention that changed our world. The wheel is the greatest invention ever made (step forward, Shiley Valentine, and take a bow – one of my favourite films (completed in 1989)). Wheels pre-date driven wheels by about 6000 years, themselves an evolution of using round logs as rollers to move a heavy load (a type of sledge). Going back in pre-history, using wheels is a practice that has not been dated. Adding wheels made any work more manageable. (If a 100 kg object is dragged for 10 m along a surface with the coefficient of friction (the reaction between two forces rubbing together) $\mu =$ 0.5, the normal force is 981 N, and the work done (required energy) is (work=force x distance) $981 \times 0.5 \times 10 = 4905$ joules. Now let us put the load on four wheels. The normal force between the four wheels and axles is the same (in total) 981 N. Assume, for wood, $\mu = 0.25$, and say the wheel diameter is 1000 mm and axle diameter is 50 mm. So while the object still moves 10 metres, the sliding frictional surfaces only slide over each other at a distance of 0.5 m. The work done is $981 \times 0.25 \times 0.5 = 123$ joules; the work done has reduced to 1/40 of that of dragging the object. I hope that makes sense; basically, using a wheelbarrow or a cart means the work can be done in a fraction of the time and effort).

The earliest wheels were made in Ancient Mesopotamia and Modern Iraq, between 3500 B.C. and 3000 B.C. There were two kinds of wheels. One was a potter's wheel used to make pottery. The other wheel was used to move heavy loads, transport, and trade items. This work would not have been possible without the axle. An axle was a rod going through the centre of the wheel, and turning the axle turned the wheel, saving both time and effort. Wheels were generally solid, but this was not to last for long. The next significant step in the evolution of the wheel was the spoked wheel, a wheel that has spokes radiating from the centre of the wheel, which made the wheel both lighter and faster.

I would now like to have a quick look through history and discover with the reader how the wheel has influenced history. I cannot cover all aspects of history; therefore, I have picked out only those pieces of history that I hope we both find interesting.

The first section deals with chariots. Egypt's armies used chariots for speedy transport on the battlefield and all-purpose war machines. The Persians added the innovation of scythed chariot wheels, long blades that stuck out from the hubs, killing enemy foot soldiers in the hundreds. Rome kept chariots for racing, hunting and ceremonies, while India used them as platforms for archers. Let us now look at some of these great Empires in history.

The first historical time frame I would like to discuss is the ancient Egyptians (3000 BC - 30BC). However, let us discuss how the wheel made the Kingdom what it became before leading to its downfall. The Old Kingdom (2686BC -2181BC) was when great monuments were constructed. Before, we mentioned the Great Pyramid of Giza and The Sphinx, built during this period. According to sources, periods of great drought led to changes in government. One excellent Pharoah characterised the New Kingdom (1549BC – 1069BC), Ramesses the Second (1303BC -1213BC), known as Ramesses the Great. There were many temples built during this time, including the twin temples of Abu Simbel. The temples took over twenty years to construct. The Great Temple originally had four giant statues of Ramesses the Great, each about 20 metres 965 ft) high. The Temple itself is thirty metres (98 ft) tall and has a width of thirty-five metres (115 ft) long. The Small Temple is about twelve metres (40 ft) high and is twenty-two (92 ft) long. This Temple is dedicated to Nefertari, the Pharaoh's wife. Six statues are adorning the entrance, four of Ramesses and two of Nefertari, ten metres (32ft) high. The temple complex was relocated in its entirety in the 1960s to avoid it being submerged during the creation of Lake Nasser, the massive artificial water reservoir formed after the building of a dam on the Nile. During Ramesses the Great's reign, there was a battle between the Hittites (a Kingdom based around modern-day Turkey and Syria) and the Egyptians called the Battle of Kadesh (around 1274 BC). I have mentioned this battle as it is likely to have been the largest chariot battle ever fought, involving over 5,000 chariots. Thanks to all the inscriptions on walls, and the survival of the Egyptian-Hittite peace treaty (the earliest peace treaty in the world), this is the best-documented battle in ancient history. I will let the reader find out more for themselves. A last note on Ramesses the Great, believed by some academics that it was during his reign that the Jews, led by Moses, left Egypt. The film, The Ten Commandments, starring Yul Brynner and Charlton Heston, made in 1956, is based on biblical records (The Book of Exodus). I will let the reader find out more for themselves. There are thousands of academic journals, papers, documents and other literature printed on the subject of the Egyptian Empire. The author believes that Egypt continues to fascinate readers to this day due to the number of buildings and writings.

The following Empire that I want to discuss is another ancient Empire, the Roman Empire. The Roman Empire was founded in 27 bc by Caesar Augustus (63BC-AD14) and existed until its fall in ad 467. The Roman Empire stretched across three continents and was in existence for over five hundred years. There has been a large amount of literature written about this time, which is beyond the scope of this book. However, in 30BC, Egypt became part of the Roman Empire after the Battle of Actium. The defeat caused the death of Cleopatra (69BC – 30BC), alongside her lover Mark Anthony (83BC -30BC). Their love affair has spawned many films, with the most famous film being Cleopatra, a 1963 film starring Richard Burton (1925-84). I will let the reader find out more for themselves. The death of Cleopatra meant that the ancient Roman Republic became the Roman Empire under Emperor (Caesar) Augustus (as mentioned). In the Roman Empire, chariots were not used for warfare, but for chariot racing, especially in circuses, or triumphal processions, when as many as ten horses could pull them. There were four divisions, or factiones, of charioteers, distinguished by the colour of their costumes: the red, blue, green and white teams. The main centre of chariot racing was the Circus Maximus, situated in the valley between the Palatine and Aventine Hills in Rome. The track could hold 12 chariots, and a raised median separated the two sides of the track termed the spina. I have mentioned the Roman Empire due to one of my top ten favourite films, Ben Hur, a film made in 1959 by MGM studios. The film won a record eleven Academy Awards. The film traces the fictional life of Judah Ben-Hur, a Judean Prince, played by Charlton Heston (1923-2008, who the Academy Award for Best Actor). His final battle will be against the Roman Tribune and sworn enemy Massala (a role that bought Stephen Boyd (1931-77) a Golden Globe for Best Supporting Actor). The battle will take the form of an epic chariot race. Of course, Ben-Hur wins! The chariot race was modelled on a Roman circus based in Jerusalem (though the actual race in the film was on a set that cost over $1 million at the time). The Vatican itself approves the film! The original book, Ben-Hur, written in 1880 by Lew Wallace, has never been out of print. The author, Lew Wallace (1827-1905), also had a fascinating life! During the Roman occupation of Britain, which was started in AD43, there were several battles with the local population. The most famous revolt against Roman occupation was led by a famous Celtic queen called Boudicea (date unknown – AD60/AD61). Boudicea was the queen of the Iceni tribe (Norfolk and Suffolk area), who led her armies into battle at Colchester (called Camulodunum during the Roman era). The troops led by Boudicea continued into London (Londinium) and Verulamium (near St Albans). Eventually, Boudicea was defeated and either took her own life or died of illness (sources vary). There is a sculpture by Thomas Thornycroft (1815-85) of Boudica and her daughters in her chariot, addressing her troops before the battle, located near Westminster Bridge. However, this battle did not involve significant numbers of chariots and was mainly fought by infantry troops.

I want to discuss the last time chariots were used in battle. Before we discuss the war itself, there are two main characters that I would like to introduce. Darius III (ruled from 336BC to 330BC) was the last Achaemenid King, a kingdom that originated in Persia (now Iran). The following person that I would like to introduce is Alexander the Great (356BC – 323BC), the King of Macedonia in Greece. This Greek king managed to create one of the ancient world's largest empires. By his death, he had held the following titles: King of Macedonia, Pharaoh of Egypt, King of Persia and King of Egypt. Alexander the Great died in Babylon at the early age of thirty-two years old. Alexander the Great named over seventy cities after himself during his campaigns in Asia and Africa. There is plenty of information available on the internet, and I hope this has whetted the reader's appetite. The invasion of Persia by Alexander the Great resulted in the Battle of Gaugamela (331 B.C.) between the Persians and Alexander's Macedonian forces. Although Alexander's forces were vastly outnumbered, it was by superior military tactics that Alexander won the battle (these tactics are still studied today). I have mentioned previously that Persians attached long swords to their chariots' hubs to attack foot soldiers in combat. When the chariots of Darius III attacked the Macedonian infantry lines, Alexander's tactic merely opened up the line and allowed the chariots to pass through and re-closed the line. The Macedonians then surrounded the Persian chariots and destroyed them. There are differing accounts as to what happened at the battle, whether Darius's army deserted him or he turned and fled. Either way, Darius was killed or murdered after the war. He was given an entire burial ceremony a Persepolis (the ceremonial capital of the Persian Empire, located sixty kilometres south of Shiraz in Iran). The grave has never been found.

After this time, more armies were employing trained cavalry; as cavalry could go where chariots could not, the chariot's heyday ended.

There are two types of sustainable power that I suggest the reader considers now. The windmill and the watermill both deserve mention. Both of these engineering feats have significantly increased the civilisation of humankind. Let us look at the windmill. By 200 BC, the Chinese were using simple windmills to pump water. By AD 600, in Persia (now Iran), windmills were used to ground grain into flour, and around AD 1100, this use was seen in Europe. The use of windmills to pump water and mill flour until 1888, when an American inventor, Charles F. Brush (1849–1929), invented the first windmill to create electricity in Cleveland, Ohio. This machine was called a wind turbine. The turbine was built in 1887–1888; it stood 60 feet high and weighed 80,000 pounds. It supported a rotor blade of 56 feet in diameter, and it stored electricity in a bank of a dozen batteries. This wind turbine used horizontal blades. A watermill is a building with equipment used for a particular purpose, such as grinding grain. The equipment consists of a wheel that is powered by moving water. There is a long history of watermills, dating back to the first century B.C. Watermills were at their peak in the nineteenth century, with around six thousand in Britain. The grain is imported into the country from the rest of the world and transported to the watermill. The watermill then grounds the grain into flour. There are several different designs of watermills, and I suggest the reader finds out more.

The last section of this chapter on wheels would typically be concentrated on the engines that power the wheels. In chapter E – Engines, I have already covered steam engines, internal combustion engines, and various cars. Therefore I will not cover these again here. I have not included many uses of the wheel in this chapter. I have deliberately left out tracked vehicles, as they can be covered in another volume. I have left out cannons and warfare in general (from the Crimean War to the Second World War and beyond), as I will also be covering these in a separate volume. I have also deliberately left out trains, as these can be covered elsewhere. However, I have tried to cover those parts of the wheel that the author has found most interesting. I have, I hope, proved that the most significant invention was the wheel.

CHAPTER TWENTY-FOUR – THE LETTERS X, Y, Z

There was only one option for the letter X: the x-ray. The x-ray is part of the electromagnetic spectrum, an all-encompassing term for all the frequencies of electromagnetic radiation. The Gamma wave has the shortest wavelength, ten picometres (one trillionth of a metre). The unit of picometre is used for measuring atoms. The next shortest wavelength is X-rays, which have a wavelength up to one nanometre (one billionth of a metre). Wavelengths visible to the human eye range from ultraviolet (100 nanometres) to infra-red (One micrometre, one-millionth of a metre). Microwaves are not visible and have a wavelength of one millimetre to one metre in length. Radio waves have a wavelength of one millimetre to a hundred kilometres. Gamma waves and X-rays can enter the human body and cause potential damage to the fabric of cells within the human body. This is called ionisation. Light, both visible and invisible and radio waves cannot enter the body and are therefore called non-ionising rays. In 1894, Wilhelm Roentgen (1845–1923) discovered that photographic plates placed near an electric charge (through Argon in a covered glass tube) were illuminated. This is called electromagnetic radiation. Eventually, he took a picture of his wife's hand using the plates, showing the skeletal structure underneath. This picture was the first X-ray, and the first Nobel Peace Prize in Physics was awarded to him in 1901. There are beneficial effects of

X-Rays. They can be used to find foreign objects in your body to treat people with cancer.

However, there is a reason why people leave the room during an X-Ray. Many academics warned that X-rays could be dangerous and cause blindness but were beneficial in finding diseases in the body. Many people, such as Dr William Dudley of Vanderbilt University, Professor John Daniel, William J Morton and Nikola Tesla, reported burns and hair loss associated with using X-Rays. Two people should be mentioned. The first is Clarence Dally (1865-1904), an assistant to Thomas Edison. From 1896, he worked with Thomas Edison producing X-ray images over four years, 1896 to 1900. These experiments subjected Dally to ever-increasing amounts of radiation. By 1900, Dally was suffering extreme effects. There was hair loss and lesions on his hands and face. Dally eventually had his left arm amputated below his shoulder and four fingers on his right hand. Eventually, the right arm had to be removed as well. Clarence Dally died in 1904 of mediastinal cancer eight years after first experimenting with X-rays and was the first man ever recorded to die of radiation poisoning. His legacy was the progression of medicine, though it was unable to help at this time. The second person was Elizabeth Fleischman (1867-1905). Elizabeth was producing X-Rays in her studio in San Francisco from 1897 onwards. Her most famous work showed a bullet lodged in the brain of Pte Gretzer, a soldier wounded in the Philippines. Her work in producing X-Rays helped the world of medicine. However, the years of unprotected effects of radiation would lead to cancer. Elizabeth Fleischman died at 38 years old and was the first woman to die due to unprotected X-Ray exposure.

There was only one option for the letter Y. I wanted to celebrate the achievements of one man. Please step forward, Brigadier General Charles Edward Yeager (1923-2020). To give him his nickname, Chuck Yeager was a United States Officer and one of the fastest men on earth at one time. Many people can look at this information on Google; I will do some highlights of this man's career for those who do not have a computer. Chuck Yeager fought in the Second World War and was credited with over eleven aircraft shot down (he got "ace in a day" for five aircraft in one day). There were other wars Chuck Yeager fought in, such as the Korean War and the Vietnam War (the reader can find out more for themselves). However, his work after the Second

World War brought him international fame. Chuck Yeager was the first man to break the Sound Barrier on 14 October 1947, flying an experimental rocket aircraft called the Bell X-1. In 1962, he joined the USAF (United States Air Force) Aerospace Research Pilot School, which trained astronauts for NASA and the Air Force. Chuck Yeager won many medals, including the Purple Heart and many Distinguished Service medals. He is still considered one of the most excellent pilots of all time.

The letter Z is the last letter of the alphabet, and there are two options that I have chosen. The first option is Zeppelins, mentioned in the first chapter. However, there is one Zeppelin that should be mentioned. The LZ-129, better known as the Hindenburg, was the largest aircraft ever built and the pride of Nazi Germany. It was 245 metres (804 feet) long and was filled with hydrogen gas instead of Helium (the Allies restricted Germany due to the Second World War). On 6 May 1937, the Hindenburg burst into flames and was destroyed whilst landing in New York. There were thirty-five people dead and sixty-three survivors. This single catastrophe marked the end of the airship era. The next option is the zodiac, a belt of constellations in the night sky. Twelve-star constellations are known as Aries, Taurus, Gemini, Cancer, Leo, Virgo, Libra, Scorpio, Sagittarius, Capricorn, Aquarius and Pisces. In later volumes, we will be exploring Space and all that it contains.

I hope that I have created enough interest in engineering to make people want to find out more. Whether this book establishes a career or a lifetime hobby, I hope to have been of service to you all.

Kind Regards,

The Author.

www.ingramcontent.com/pod-product-compliance
Lightning Source LLC
Chambersburg PA
CBHW051315220526
45468CB00004B/1352